AGRICULTURE ISSUES AND POLICIES

U.S. FOOD MARKETING: ISSUES AND CHANGES

AGRICULTURE ISSUES AND POLICIES

Additional books in this series can be found on Nova's website under the Series tab.

Additional E-books in this series can be found on Nova's website under the E-book tab.

FOOD AND BEVERAGE CONSUMPTION AND HEALTH

Additional books in this series can be found on Nova's website under the Series tab.

Additional E-books in this series can be found on Nova's website under the E-book tab.

AGRICULTURE ISSUES AND POLICIES

U.S. FOOD MARKETING : ISSUES AND CHANGES

WALTER J. KINKOFF
EDITOR

Nova Science Publishers, Inc.
New York

LIBRARY OF CONGRESS CATALOGING-IN-PUBLICATION DATA

U.S. food marketing : issues and changes / editor: Walter J. Kinkoff.
p. cm.
United States food marketing
Includes index.
ISBN 978-1-61761-731-7 (hardcover)
1. Food industry and trade--United States. 2. Food--United States--Marketing. 3. Organic farming--Economic aspects--United States. I. Kinkoff, Walter J. II. Title: United States food marketing.
HD9005.U233 2011
381'.410973--dc22

2010033092

Published by Nova Science Publishers, Inc. † New York

CONTENTS

Preface **vii**

Chapter 1 Supply Chain Basics: The Dynamics of Change in the U.S. Food Marketing Environment **1**
United States Department of Agriculture

Chapter 2 Breaking Down Market Barriers for Small and Mid-Sized Organic Growers **55**
Alida Cantor and Ron Strochlic

Chapter 3 The U.S. Food Marketing System: Recent Developments, 1997-2006 **111**
Steve W. Martinez

Chapter Sources **175**

Index **177**

PREFACE

Smaller niche-market food producers have good reason to be optimistic in today's marketplace. Increased demand for specialty food products, and consumers' willingness to visit different retailers to get them, is creating new marketing opportunities for food producers and processors that can offer innovative merchandise designed to meet the specific needs and preferences of particular consumer segments. Several of today's most highly profitable retail supermarket chains, typically, privately held firms that operate relatively small numbers of stores, have succeeded because of their ability to closely match their food product offerings to the specific needs of their customer base. Such firms offer promising marketing opportunities to small and medium sized food suppliers who can meet their basic product requirements. This book provides an overview of the changing retail landscape and identifies some of the characteristics associated with successful food retailing

Chapter 1- The growing retail desire for exclusive and specialized food products offers new marketing opportunities for small and medium-sized food suppliers that understand the new world of food merchandising and are capable of delivering food products that satisfy commercial requirements for quality, innovation, and value. This document provides an overview of the changing retail landscape and identifies some of the characteristics associated with successful food retailing. It is designed to help smaller scale food producers and processors develop profiTable business strategies and identify customers likely to appreciate their unique products.

Chapter 2- We are very grateful to the many individuals and organizations that have worked to make this research possible, particularly the USDA Agricultural Marketing Service for providing funding through Cooperative

Agreement 12-25-A-4823 and Adam Diamond of the USDA Agricultural Marketing Service for his assistance with all aspects of the research.

Many thanks to Paige Culver, CIRS intern, who provided invaluable support with her comprehensive analysis and write- up of the survey results. We are very grateful to Lisa Kresge, CIRS research associate, for her support in survey design and editing, and to Gail Wadsworth and Kate Snow for their editorial assistance. We would also like to thank Margaret MacSems for assistance conducting interviews with Spanish-speaking growers.

Chapter 3- A main development in the U.S. food marketing system is the influx of stores that have not traditionally sold food but that increasingly offer food items. With the growing trend for eating out, traditional food retailers also face competition from the foodservice industry.

In: U.S. Food Marketing: Issues and Changes ISBN: 978-1-61761-731-7
Editors: Walter J. Kinkoff

Chapter 1

SUPPLY CHAIN BASICS: THE DYNAMICS OF CHANGE IN THE U.S. FOOD MARKETING ENVIRONMENT

United States Department of Agriculture

NEW WAYS IN FOOD MARKETING

The growing retail desire for exclusive and specialized food products offers new marketing opportunities for small and medium-sized food suppliers that understand the new world of food merchandising and are capable of delivering food products that satisfy commercial requirements for quality, innovation, and value. This document provides an overview of the changing retail landscape and identifies some of the characteristics associated with successful food retailing. It is designed to help smaller scale food producers and processors develop profiTable business strategies and identify customers likely to appreciate their unique products.

Smaller niche-market food producers have good reason to be optimistic in today's marketplace. Increased demand for specialty food products—and consumers' willingness to visit different retailers to get them—is creating new marketing opportunities for food producers and processors that can offer innovative merchandise designed to meet the specific needs and preferences of particular consumer segments. Several of today's most highly profiTable retail supermarket chains—typically, privately held firms that operate relatively small numbers of stores—have succeeded because of their ability to closely match their food product offerings to the specific needs of their customer base. Such firms offer promising marketing opportunities to small- and medium-sized food suppliers who can meet their basic product requirements.

Furthermore, the nature of business transactions across the U.S. food supply chain is rapidly being transformed. A natural competitive tension between food buyers and sellers that characterized business relationships appears to be evolving toward a new paradigm of cooperation, encouraging retail businesses to reconsider their traditional roles and adjust their practices to accommodate this changing business climate. Brian Silbermann, president of the Produce Marketing Association, notes that only a few years ago:

- Traditional full-service supermarkets dominated the U.S. food retail landscape.
- Shippers sold to many buyers in wholesale and retail markets, and no single large buyers dominated the market.
- Retail buyers were transaction- and price-oriented, and not open to cooperative partnerships with suppliers.[1]

In today's marketing environment, however, many retail and foodservice channels compete for the consumer food dollar, and increasingly affluent and discriminating shoppers—no longer content to rely on their neighborhood supermarkets for the bulk of their grocery items—seek out products that appeal to their health, environmental, and social interests, and that suit their time-pressed lifestyle. As a result:

- The importance of national brands has receded as food retailers promote their firm's reputation through private label programs and other unique offerings.
- Retail food buyers seek to establish longer term relationships with fewer suppliers to ensure steadier pricing, supply availability, and better quality control.
- The cultural divide between the procurement and marketing sides of food retail firms is disappearing as long-term supply relationships take precedence over short-term price negotiations.

The rapid shift toward supply chain integration is also being encouraged by the growing sophistication and declining cost of technology, especially wireless technology, which enables food suppliers and retail buyers to exchange increasingly elaborate data on a cost-effective basis, and to more precisely match the availability of supplies and movement of inventory with demand requirements at all stages of distribution and marketing. (For more information, see Global Data Synchronization on page 27.)

What do these supply chain developments mean for the smaller scale producer, processor, or agricultural cooperative that wishes to sell food products to retail grocers? As the number of retail accounts declines through consolidation (see Retail Concentration on page 24) and retailers become more selective about their business partners, food suppliers must increasingly focus on producing and delivering merchandise that:

- Appeals to growing consumer desire for quality, variety, and innovation.
- Addresses the emerging needs of Asian, Hispanic, and African-American consumers.
- Helps retailers tell their customers the origin of food and the production methods used to raise it.
- Enables retailers to differentiate themselves from their competition, as with locally branded products and exclusive private label items.

Moreover, to be considered as vendors, food producers and processors must:

- Comply with the stringent requirements of a preferred vendor program or production contract in their product preparation, packaging, and quality.
- Be able to participate in real-time electronic data exchange with retail customers using compatible software.
- Use appropriate technologies, such as barcodes and radio frequency identification (RFID), to maintain quality control, enhance transaction efficiency, and reduce recording errors.

This publication addresses two major trends changing retail food marketing—a move toward differentiation as a marketing strategy and a simultaneous shift towards vertical integration between food suppliers and buyers. It examines the ramifications of these developments for the smaller scale food supplier and identifies strategies for remaining competitive in this environment.

Table 1. Sales of Food at Home by Type of Outlet, in Percent

Outlet Type	1986	1996	2006
Conventional supermarkets	65.3	58.2	57.7
Other grocery	14.3	15.8	3.4
Specialty food stores	5.8	2.6	2.7
Supercenters and wholesale club stores	**0.4**	**4.3**	**17.9**
Mass merchandise stores	1.5	2.1	1.7
Convenience stores	3.3	3.0	2.9
Home deliveries/mail order	1.2	2.8	3.8
Farmers/processors/ wholesalers	2.0	1.9	0.8
Other stores	6.2	10.1	9.2

Source: USDA/ERS

RETAIL DIFFERENTIATION

Traditional neighborhood supermarkets are losing their dominance of the grocery trade among consumers, especially for nonperishable items. The erosion of consumer loyalty to national brands has led buyers to patronize supercenters and discount outlets such as dollar stores, while competition from alternative retail outlets, such as natural food markets, ethnic food markets, and direct marketing venues, has further nipped away at the share of the perishables market long dominated by the supermarket sector. The one-stop shopping formula that worked so successfully for supermarkets in the past is giving way to an increasingly fragmented and diverse retail market, in which consumers are willing to seek out and patronize different outlets to satisfy their needs for value, quality, and variety. As noted by Deloitte Touche Tohmatsu's research department in March 2003:

> The differences in product and shopping preferences between young and old, rich and poor, married and single, Caucasian and ethnic, urban and exurban, are becoming greater. These consumers are looking for products and services authentic and unique to their communities. One-size-fits-all retailing just will no longer work for a growing segment of the population.[2]

Geographic proximity no longer guarantees as table customer base for food retailers. Today's consumers are willing to shop at several stores to satisfy their grocery needs. According to a 2005 study by the Food Marketing

Institute (FMI), the average household grocery shopper visited more than one store each week to shop for groceries.[3] Fewer than half (46 percent) relied almost exclusively on a full-service supermarket for their groceries. Almost one in five (19 percent) regularly purchased grocery items at supercenters and 12 percent regularly shopped for grocery items at discount stores.

Data tracked by USDA's Economic Research Service (ERS), based on ACNielsen Fresh Foods Homescan data, reveal similar trends. By 2006, only a slight majority of U.S. food expenditures took place at conventional supermarkets, compared with nearly two-thirds of food expenditures in 1986 (Table 1). The greatest change in U.S. food shopper behavior is the extent to which food shoppers now rely on nonsupermarket stores as a source of grocery supplies. Compared with a minuscule 0.4% share of U.S. food sales for at-home consumption in 1986, the share of food sales for at-home consumption that took place at supercenters and wholesale club stores accounted for nearly 18 percent of all such sales in 2006. In the words of supermarket industry consultant Phil Lempert, the increasing willingness of the average U.S. shopper to patronize different food stores demonstrates that many supermarkets are "missing an opportunity to capture more consumer dollars by delivering and communicating . . . greater selection and value to their shoppers."[4]

Consequently, while conventional supermarkets remain the primary outlet for many high-value and perishable food products, they no longer supply the majority of nonprescription drugs, paper, household cleaning, or pet products to consumers. Supercenters now occupy an equal share of the market in many of these product categories. Warehouse club stores have also become an increasingly important outlet for grocery products, and are currently the third most important outlet for meat or poultry, breakfast cereal, and frozen foods.[5]

THE CHANGING FACE OF FOOD SHOPPING

Conventional supermarkets are store formats offering a full line of groceries, meat, and produce at average prices. They are up to 30,000 square feet in size and have at least $2 million in annual sales.

Mass merchandisers (Target, Wal-Mart) are stores that primarily sell household items, electronic goods, and clothing, but also offer packaged food products.

Supercenters (SuperTarget, Wal-Mart Supercenter) are large food, drug, and mass merchandise combination stores operating under one roof. They average more than 170,000 square feet, offer a variety of food and nonfood merchandise, and typically devote as much as 40 percent of their space to grocery items.

Warehouse stores (Food 4 Less, Super Saver) are retail/wholesale hybrids with a limited variety of products presented in a warehouse-type environment. Most items are sold at deep discounts in large packages or bulk volumes.

Wholesale clubs (Costco, Sam's Club) are warehouse stores that sell only to members. Memberships require an annual fee.

Super warehouse stores (Cub Foods, Shoppers Food & Pharmacy) are high-volume hybrids of a large traditional supermarket and a warehouse store; they typically offer a full range of service departments, quality perishables, and reduced prices.

Specialty food stores offer a more limited selection than conventional supermarkets, often specializing in organic/natural food, ethnic food, or gourmet food.

Convenience stores (7-Eleven, Quick Stop) are small variety stores with a limited selection of popular items. They usually sell items at higher prices than other stores, but offer the advantage of convenience because they are located either in residential neighborhoods or at gas stations.

Dollar stores (Dollar General, Dollar Tree, Family Dollar) have small store formats and offer food and consumable items at aggressive prices, typically in units or multiples of one dollar.

Fresh format stores (Whole Foods) emphasize perishables and offer center-store assortments—dry grocery, dairy, and frozen foods displayed in the inner aisles of the store—that differ from those of traditional retailers, especially in the area of ethnic, natural, and organic foods.

Limited-assortment stores (Aldi, Sav-A-Lot, Trader Joe's) are discount-priced grocery stores that offer a limited assortment of center-store and perishable items.

Other grocery stores generally refers to small neighborhood stores.

Trends in Retail Differentiation

How are mainstream food retailers responding to these changing consumer preferences and shopping patterns? They are differentiating themselves in two ways: by offering competitively priced private label products that meet their quality specifications and promote their firm's image, and by featuring products that target specific consumer market segments and appeal to consumers' desire for health, nutrition, variety, and convenience.

Today's food retailers are also able to take advantage of the growing capacity of technology, which enables precise monitoring of customer preferences and supply replenishment needs and permits delivery of services and products tailored to household consumers. (For more about the growing impact of technology on retail practices, see Global Data Synchronization on page 27.)

Decline of the Traditional Full-Service Supermarket

Conventional supermarkets and grocery stores have experienced steady declines in consumer demand every year since 1995, when ACNielsen began its annual survey of household shopping patterns. Between 1995 and 2004, the number of trips made by the average household each year to a supermarket or grocery store fell 25 percent, from 92 to 69 trips, and a small percentage of household consumers had stopped patronizing traditional supermarkets or grocery stores entirely by 2004 (Table 2).

Much of the decline in supermarket and grocery store patronage can be attributed to a rise in consumer shopping at discount stores such as supercenters and dollar stores. The average number of visits to supercenters for groceries rose from 15 a year in 1999 to 27 a year in 2004. At the same time, dollar stores increased their penetration; more than two-thirds of surveyed households (67 percent) visited a dollar store in 2004 compared with a bare majority of households (52 percent) in 1999. Younger consumers especially are buying groceries at supercenters and discount outlets. Twenty-nine percent of survey respondents age 15–24 used a supercenter or discount outlet as their primary food store in 2004.[6]

No Slowdown in Sight for Expansion of Alternative Retail Channels

With consumers turning away from traditional supermarkets and grocery stores as primary food sources, the importance of the retail grocery channel has declined—and the decline shows no sign of abating. By 2011, non-traditional food retailers are expected to account for 40 percent of the Nation's sales of groceries and consumables, almost equal to the 43.5 percent share expected from traditional supermarkets and groceries (Table 3).

Table 2. Changes in Food Store Patronage among U.S. Households, 1999-2004

	Percent of Households Using Channel					
Channel	**1999**	**2000**	**2001**	**2002**	**2003**	**2004**
Grocery store	**100**	**100**	**100**	**100**	**100**	**99**
Mass merchandise store	95	94	93	91	91	89
Drug store	87	86	86	86	85	84
Supercenter store	52	54	53	55	54	54
Dollar store	52	55	59	62	66	67
Warehouse store	50	49	50	52	51	51
Convenience store/Gas station	50	48	45	46	45	44
	Trips Per Year					
Channel	**1999**	**2000**	**2001**	**2002**	**2003**	**2004**
Grocery store	**83**	**78**	**75**	**73**	**72**	**69**
Mass merchandise store	26	25	23	22	21	20
Drug store	15	15	15	15	15	15
Supercenter store	**15**	**17**	**18**	**21**	**25**	**27**
Dollar store	10	10	11	12	13	13
Warehouse store	9	10	10	10	11	11
Convenience store/Gas station	13	14	15	14	15	15

Source. "U.S. Consumers Continue Trend of Shopping Less Often in Traditional Grocery Stores," ACNielsen, January 18, 2005.

Table 3. Projected Market Share, Grocery and Consumables, 2011

Retail Channel Format	Number of Stores	Annual Sales ($millions)	Dollar Share (in percent)
Traditional Grocery Stores	**40,831**	**$446,024**	**43.5**
Conventional supermarkets	27,115	382,226	37.3
Fresh format stores	981	9,815	1.0
Limited-assortment stores	3,844	25,833	2.5
Super warehouse stores	485	16,380	1.6
Other (small grocery stores)	8,405	11,770	1.1
Convenience Stores	**148,131**	**$166,094**	**16.2**
Convenience stores with gas stations	119,662	143,742	14.0
Convenience stores without gas stations	28,469	22,352	2.2
Non-Traditional Grocery Stores	**58,074**	**$413,923**	**40.3**
Wholesale club stores	1,441	84,173	8.2
Supercenters	3,264	207,958	20.3
Dollar stores	27,696	20,607	2.0
Drug stores	22,097	48,968	4.8
Mass merchandise stores	3,407	47,661	4.6
Military commissaries	170	4,555	0.4
Total All Formats	**247,036**	**$1,026,041**	**100.0**

Source. "The Future of Food Retailing," June 2007, Willard Bishop Consulting, Ltd.

Away-From-Home Food Consumption Increases at Expense of At-Home Food Preparation

The gradual rise in away-from-home food consumption is further eroding the importance of traditional groceries in the United States, leaving supermarkets and nontraditional food retailers competing for a dwindling pool of consumer food dollars. In 2006, consumers spent approximately $1.1 trillion on food consumed in the United States. Foodservice facilities supplied about $529 billion, or 48.9 percent, of this total for on-premise or immediate consumption, the third highest level in U.S. history (figure 1).

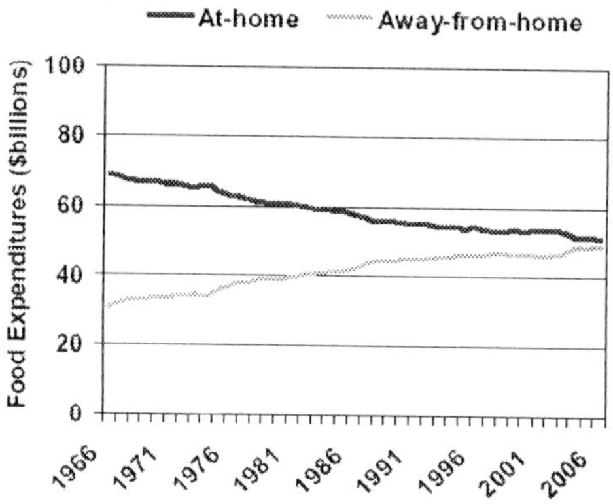

Source. Food CPI, Prices and Expenditures: Food and Alcoholic Beverages: Total Expenditures, ERS.

Figure 1. Trends in U.S. consumer expenditures for food, 1966-2006

The tendency to spend money on food outside the home is largely inversely related to age, with the youngest consumers in the United States more likely to spend their food dollars on food consumed outside the home than any other age cohort. Consumers under 25 spend more than 50 percent of their food dollars outside the home, compared with 34 percent among consumers age 75 or greater (figure 2).

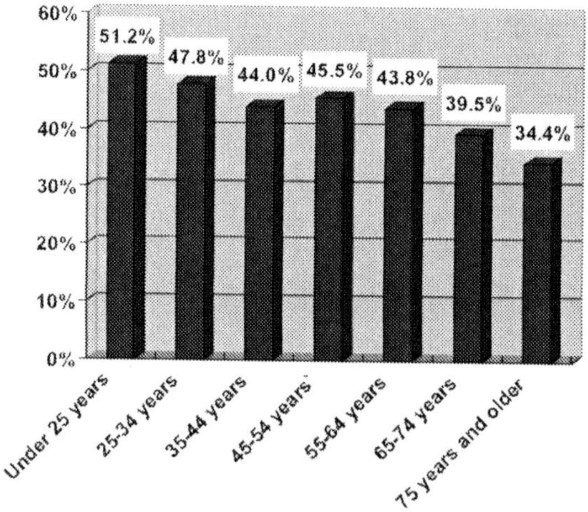

Source. Consumer Expenditures in 2005, Bureau of Labor Statistics, U.S. Department of Labor, April 2007.

Figure 2. Expenditures for food consumed away from home, by age group

Direct-to-Consumer Marketing Outlets Gaining Traction

The popularity of direct-to-consumer marketing outlets, such as farmers markets, roadside stands, and community-supported agriculture, is another challenge to supermarket profitability. According to the 2002 Census of Agriculture, the value of direct-to-consumer food sales in the United States grew 37 percent between 1997 and 2002—from $592 million to $812 million—refl ecting the enormous growth in the number and accessibility of direct-to-consumer marketing outlets, especially in urban and suburban neighborhoods. This increase has been bolstered by the growth in the number of farmers markets in the United States—from around 1,750 in 1994 to approximately 4,475 by late 2007.[7] By the end of 2005, farmers markets in the United States were estimated to generate more than $1 billion in sales per year.[8] Furthermore, the number of community-supported agriculture (CSA) operations, in which customers purchase advance shares of a farm's production in return for regular deliveries during the growing season, has expanded from an estimated 60 operations in 1990 to approximately 1,150 operations in early 2007.[9]

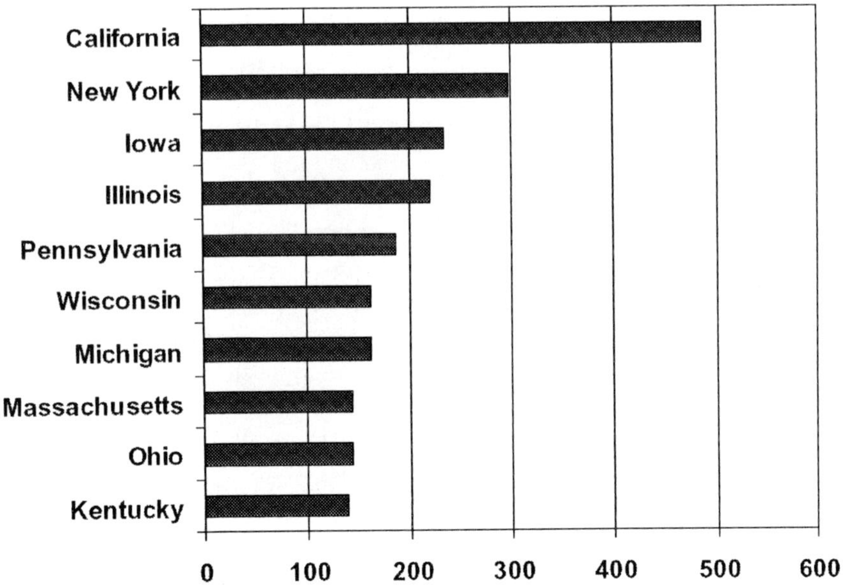

Source. USDA Agricultural Marketing Service, September 2007

Figure 3. States with the most farmers markets

The farmers market movement in the United States was initially concentrated along the West Coast and within parts of the Northeast. However, during the last few years the number of farmers markets throughout the country has increased significantly, especially in Midwestern States with a tradition of small farms, and where farms are close to population centers (figure 3). Direct-to-consumer marketing channels appear to be especially important to buyers of organic food products. According to the Organic Trade Association, about 7 percent of U.S. organic food sales in 2005 occurred through direct sales at farmers markets and other nonretail direct market outlets (including sales to foodservice customers).[10] In contrast, only 3.9 percent of all U.S. food sales in 2005 were made through any form of direct sale or home/mail order delivery.[11]

Why Consumers Love Buying Directly From Farmers

Permits them to:

- Obtain seasonal foods and other farm products at peak condition.
- Have access to varieties of farm products grown or raised for flavor, rather than tolerance of long-distance shipping.
- Receive fresher products with a longer shelf life.
- Know more about the origin of food and the methods used to produce it.
- Develop a personal relationship with local vendors, and receive more individualized or more responsive customer service.
- Potentially contribute to a reduction in fossil fuel usage by substituting locally grown food for food shipped over long distances.
- Provide additional income to family-run farms and the local economy.
- Help preserve local and regional culinary traditions.

The exponential growth in access to direct marketing outlets, such as farmers markets and CSAs, throughout the country appears to have stimulated significant growth in the value of locally grown food sold to consumers. The market research firm Packaged Facts estimated in 2007 that consumer demand for locally grown food could rise from around $4 billion in 2002 to $5 billion a year by the end of 2007 and to as much as $7 billion a year by 2012, reflecting both increased consumer patronage of farmers markets and CSAs and an expanded effort by retail and foodservice firms to procure more locally grown food that appeals to consumers.[12] Indeed, the popularity of farmers markets has grown so rapidly in the United States that a recent national survey reports that 2 percent of U.S. food shoppers now say farmers markets are their primary food shopping venue.[13] In light of these developments in the marketplace, mainstream news media are giving increasingly prominent attention to the growth of the local food movement.

Dwindling Store Loyalty Translates Into Lower Supermarket Profits

As might be expected, the steady decline in supermarket and grocery store patronage—infl uenced by the growing popularity of alternative retail channels and away-from-home food consumption—has contributed to sluggish sales growth in the traditional grocery sector. According to U.S. Census Bureau statistics, overall sales growth at conventional supermarkets and grocery stores averaged 2.4 percent a year between 1995 and 2005, compared with 16.6 percent a year at supercenters and warehouse club stores during the same period (figure 4).

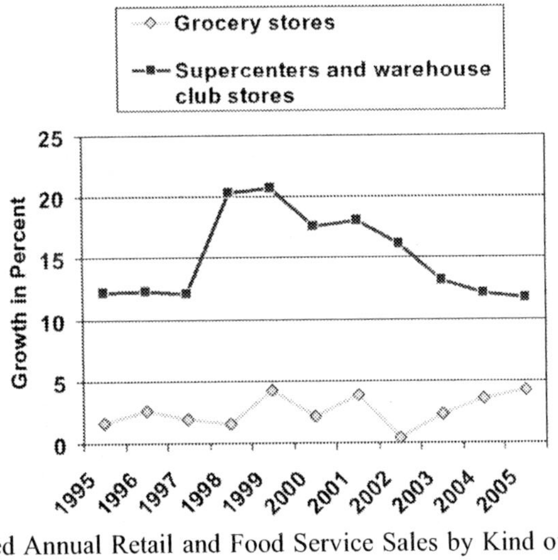

Source. "Estimated Annual Retail and Food Service Sales by Kind of Business, 1992 Through 2005," Annual Benchmark Report for Retail Trade and Food Services: January 1992 Through February 2006, U.S. Census Bureau, March 2006.

Figure 4. Annual retail sales growth, grocery stores versus supercenters and warehouse club stores, 1995-2005

Furthermore, when deflated by the food price index, it appears that recent sales growth in the U.S. supermarket sector has struggled to keep up with the rate of U.S. population growth, which currently averages just under 0.9 percent per year.[14] Although U.S. supermarket sales rose 4.6 percent in 2006 over 2005 levels, the industry's highest growth rate in 6 years, the Food Institute reported that "real" growth was actually 1.82 percent once inflation was taken into account.[15, 16] These figures suggest that even under unusually favorable circumstances, the growth of the supermarket industry is barely keeping pace with the growth in U.S. population.

Nor does this 2006 sales growth figure reflect the impact of hefty food price inflation during 2007, accentuated by energy price hikes and competing demands for agricultural commodities. By August 2007, the Consumer Price Index (CPI) for food had already risen 4.3 percent above the comparable level in 2006, and the ERS predicted that, by the end of 2007, the food CPI will have risen 3.5 to 4.5 percent over the previous year's level.[17] Moreover, the steepest rise in food prices is expected to take place among food items consumed at home, marking a significant departure from recent trends. Food-at-home prices are forecast to increase 3.5 to 4.5 percent between 2006 and 2007, well above the rise of 1.7 percent in food-at-home prices between 2005 and 2006.[18] In view of these developments, it's not surprising that many food retailers are expected to face sluggish growth over the next few years, with traditional supermarkets continuing to lag well behind food retailers with highly specialized formats, such as fresh format stores, limited assortment stores, and supercenters (figure 5).

Succeeding in the "Brave New World" of Food Marketing

Although the near-term prospects for the U.S. supermarket industry seem somewhat bleak overall, some individual retailers have fared extremely well—a phenomenon easily obscured by looking at average sales alone. In 2004, the top 25 percent of performers in the supermarket industry increased sales by more than 7 percent and the top 50 percent of performers posted sales gains of nearly 5 percent. In contrast, the average sales gain for the bottom 25 percent of supermarket firms was only 0.5 percent.[19]

Many supermarket chains that seem to be staying "ahead of the curve" are those that successfully differentiate themselves from their competitors by offering products and services that appeal to specific segments of consumers, especially if they adopt a cautious approach to store expansion and financial leverage. These retail businesses are astutely responding to a general movement within the United States toward social fragmentation and deconsolidation. According to one industry analyst, in "virtually every [. . .] aspect of society, polarization is occurring, made manifest in various ways:

- The widely divergent voting blocs in the Nation
- A divergence of lifestyles between married and single people
- Geographical choices—leaving suburbs for cities on one hand, and suburbs for exurbs on the other
- A growing distinction between affluent and impoverished
- An increasing separation of young from old
- An increasing separation of ethnic groups from each other, and from the mainstream
- An increasing fragmentation of information-delivery means."[20]

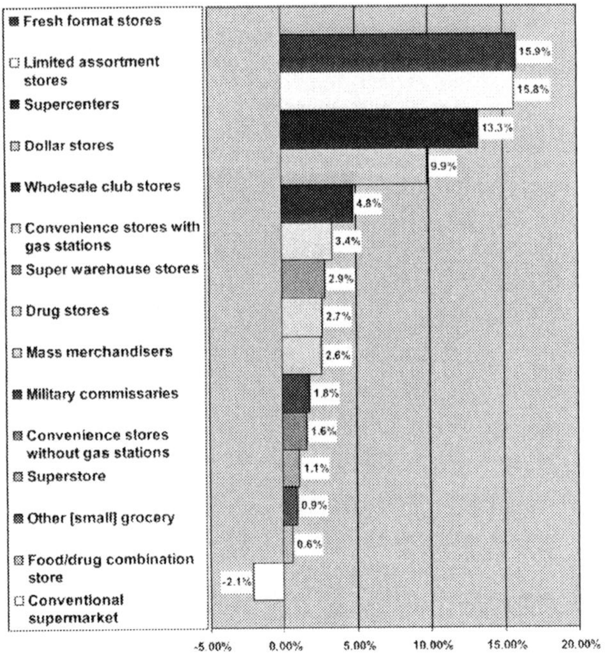

Source. "Sales Growth for Grocery Retailers Not Expected to Keep Pace with Inflation," Food Institute, <www.foodinstitute.com/ pressreleases.cfm>. Projections based on Willard Bishop and ERS forecasts.

Figure 5. Expected compound annual sales growth rate: 2005-2010 (grocery and consumables)

Bigger is Not Always Better

Supermarket chains with a relatively small number of stores can compete successfully with larger chains by using innovative merchandising techniques and building reputations for quality and service. For example:

- In 2004, Raley's and Wegmans Food Market ranked among the largest 25 U.S. food retailers in annual sales. This ranking is especially impressive when you consider that Raley's operated only 137 stores and Wegmans Food Market 67 stores at the time of the report, compared with an average of 853 stores among other supermarket firms listed in the top 25.
- By 2004, Whole Foods Market, a national retailer of natural and organic foods, had climbed into the top 20 list of largest U.S. food retailers, with annual sales of $3.9 billion, even though the firm operated only 166 retail stores at the time. Chief Executive Officer John Mackey reported that 2004 had been the firm's best year to date, with a 14.9-percent increase in same-store sales growth contributing to an overall 23-percent increase in sales over 2003 levels.
- In 2006, Wegmans Food Market, Publix, and Raley's—all privately held companies—were rated by Consumer Reports magazine the best conventional supermarket chains in the United States.[24]

Another indicator that "bigger is not always better" is that supermarket firms with relatively small annual sales volumes are doing better financially than many of their larger peers. In FY 2003, for example, food retailers with annual sales under $100 million posted 6-year highs in net profits (1.45 percent) and return on equity (20.38 percent), compared with an average net profit for the industry of 0.88 percent, and a return on equity of 9.4 percent.[25]

Importance of Target Marketing

Recent figures on new supermarket store construction reveal the growing importance of target marketing among supermarket firms. Stores catering specifically to consumers of ethnic, natural, and high-end "gourmet" foods account for a growing percentage of food retail outlets. According to the FMI, 14.5 percent of surveyed food retail firms opened at least one niche-focused store in 2005. Of these, 44 percent had ethnic formats, predominantly oriented toward the Hispanic consumer market. Retailers were also building natural and organic stores for consumers concerned about health and wellness and gourmet outlets for high-income shoppers.[26]

As stocking a broad range of products gives way to increased emphasis on target marketing, the average size of new supermarkets in the United States also has declined substantially in recent years (figure 6). The average size of new supermarkets fell from 47,500 square feet in 2002 to 34,000 square feet in 2003, the first time in 10 years that it had fallen below 40,000 square feet.

Given expectations for flat growth in the traditional grocery sector during the next few years, it comes as little surprise that food retailers are attempting to capture market share in the few sectors with prospects for expansion: organic foods, gourmet foods, and various ethnic foods. A brief glance at prospects in each sector reveals why consumers of organic/natural, gourmet, and ethnic foods attract such disproportionate attention from U.S. food retailers.

Growth of Organic Food Markets

Despite several consecutive years of aggressive sales growth in recent years (figure 7), the organic food market is expected to continue expanding rapidly, rising an average of 16 percent per year between 2005 and 2010, compared with a projected annual growth of 2–3 percent in the conventional food sector.[27] With organic food sales currently representing less than 3 percent of the U.S. food market, consumer demand for organic foods appears far from saturated.[28] A slight slowdown in growth rates for well-established product categories such as fresh produce (about 39 percent of the organic food market) is likely to be overshadowed by a growth surge in such less-well-established product categories as meat, poultry, dairy, and sauces/ condiments.[29]

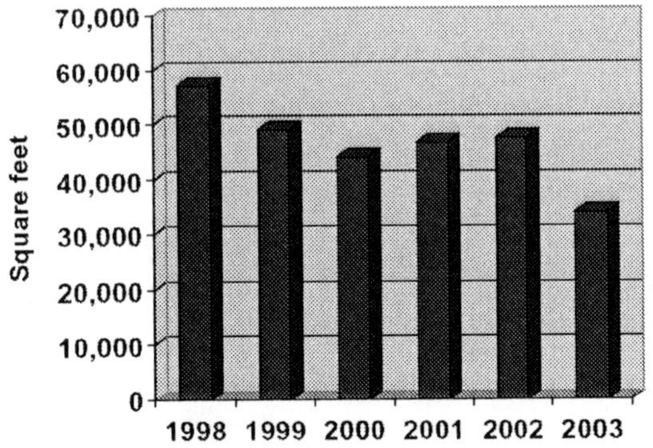

Source. FMI

Figure 6. Average size of new supermarket stores in the United States

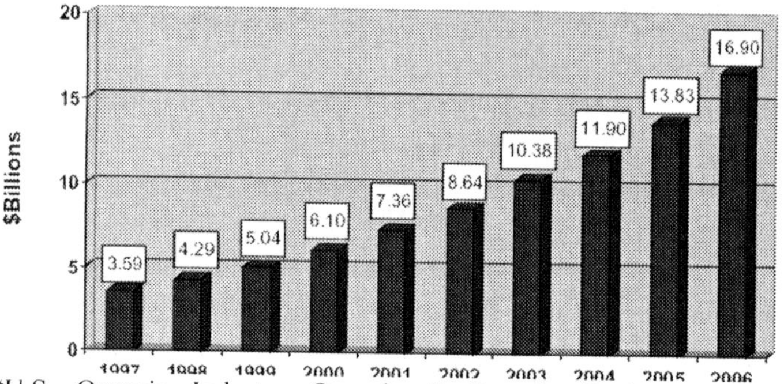

Source. "U.S. Organic Industry Overview," Organic Trade Association 2006 Manufacturer Survey. <http://www.ota.com/pics/documents/short%20 overview% 20MMS. pdf>

Figure 7. Annual growth in U.S. organic food sales, 1997–2006

The Hartman Group, a market research firm that investigates trends in the natural product marketplace, reported that two-thirds of U.S. consumers purchased organic food and beverages on an occasional basis in 2004 and that 30 percent purchased them on a regular (daily or weekly) basis.[30] Additionally, organic food purchases were motivated primarily by health and nutritional concerns, most specifically an interest in avoiding pesticides, chemicals, antibiotics, and growth hormones. However, the motivation behind organic food and beverage purchases becomes more complex as the level of organic patronage increases. Among the most dedicated "core" organic consumers (accounting for 21 percent of organic food and beverage shoppers), most prefer to support brands that "are not owned by conglomerates" and that have "an authentic story behind their creation."[31] The interest of core organic consumers in learning where their food comes from and in supporting small businesses creates unique opportunities for those food producers and processors that can satisfy this interest.

In addition, the organic food market can expect to receive a boost from demographic trends as increased buying power among Hispanic-Americans, Asian-Americans, and African-Americans should further help promote consumer demand.[32] Recent analysis of organic food patronage by the Hartman Group indicates that Caucasian consumers in the United States as a whole are less likely than other racial/ethnic groups to purchase organic foods on a regular basis.

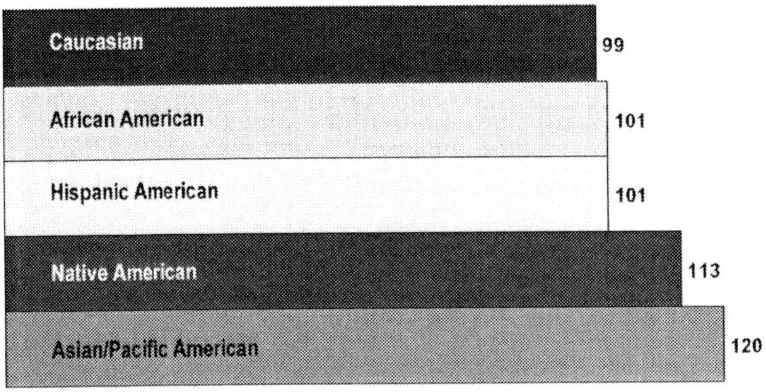

Source. Organic Food & Beverage Trends 2004: Lifestyle, Language and Category Adoption," The Hartman Group, Summer 2004, pp. 26–27.

Figure 8. Organic ethnic purchase index

Growth of Gourmet Markets

According to the market research firm Packaged Facts, the U.S. market for high-end "gourmet" foods and beverages is currently worth $41.2 billion and can be expected to reach $62 billion by 2009.[33] Nearly 20 percent of surveyed food shoppers report they try to eat gourmet food "whenever they can."[34] Several factors contribute to the growing demand for gourmet foods, including a greater commitment to the category on the part of mainstream food marketers, the expanding availability of gourmet food products (especially those in convenient, ready-to-eat formats), and a growing synergy between the natural and gourmet foods industries. From Packaged Facts:

> More gourmet products are using natural and organic ingredients (and appealing to consumers' social consciences by offering products that are good for the environment and for the workers who produce them), and more natural foods retailers are carrying upscale, gourmet food and beverage items that meet their quality standards.[35]

Changing Ethnic Demographics

Another development triggering profound changes in grocery merchandising is the expanding contribution of Hispanic, Asian, and African-American consumers to the market, refl ecting geographic shifts, population growth, and increasing affluence among these segments of the consumer population. As the United States becomes more ethnically diverse, the relative importance of Hispanic, Asian, and African-American purchasing power has grown enormously, changing the nature of consumer demand in many localities. Within the 5-year period 1999–2004, the buying power, or total amount of personal income available for spending on goods and services by Hispanic, Asian, and African-American residents increased by several hundred percent in many States[36] (figures 9-11):

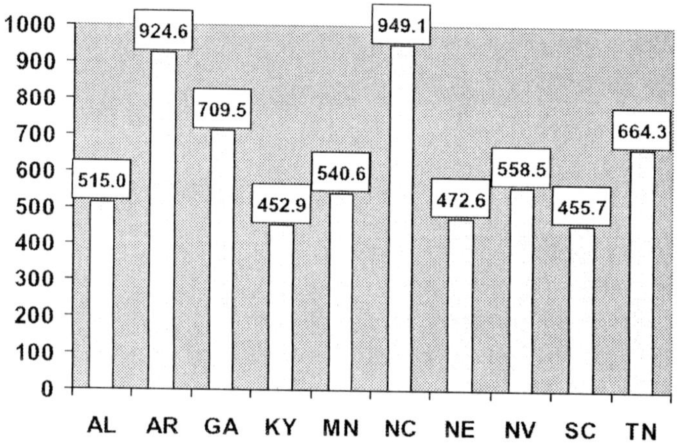

Figure 9. Percentage increase in Hispanic buying power, by State, 1999-2004 (ten fastest growing States)

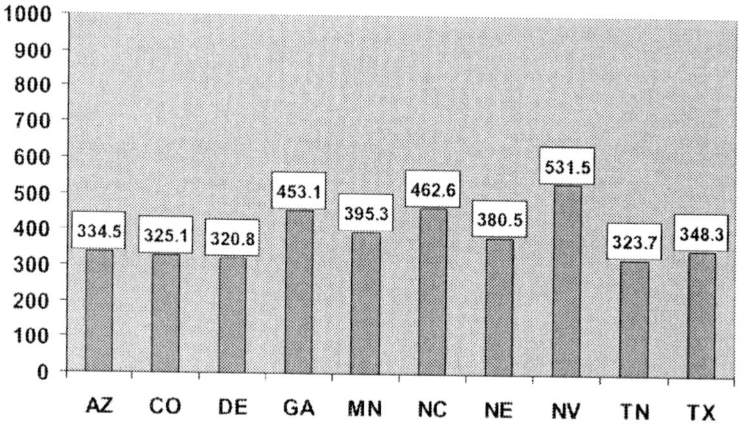

Figure 10. Percentage increase in Asian buying power, by State, 1999-2004 (ten fastest growing States)

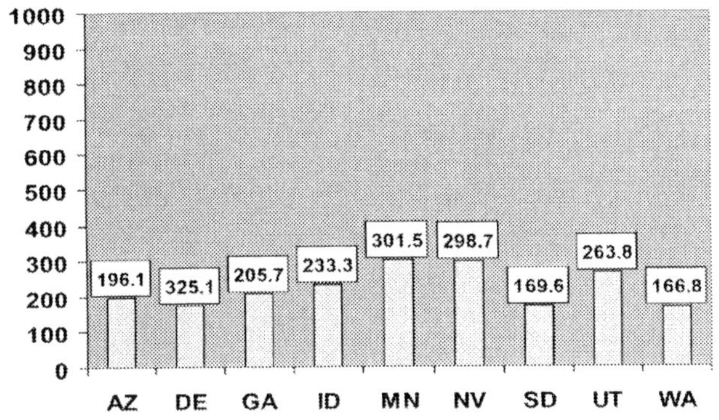

Figure 11. Percentage increase in African-American buying power, by State, 1999-2004 (ten fastest growing States)

New Market Opportunities

In many ways, today's retail marketplace offers great opportunities for a smaller food producer able to supply steady volumes of high-quality, consumer-ready products to retail markets. To stay competitive in today's highly concentrated grocery sector, food retailers are offering an expanded array of highly differentiated and customized products to meet the needs of their increasingly diverse and discriminating customer base. The average number of stock keeping units (SKUs) in produce departments, for example, more than doubled between 1980 and 2006 (figure 12).[37]

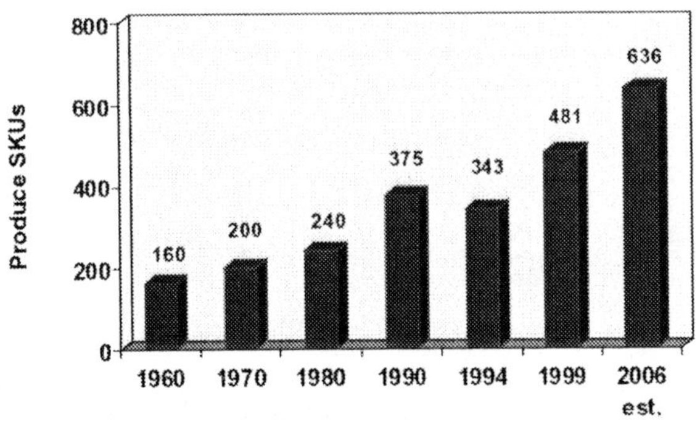

Source. FreshTrack 2001, Cornell University

Figure 12. Average number of produce SKUs offered by retail outlets

Promoting Local Origin: Case Study #1: Appalachian Harvest

Two regional supermarket chains that have been leaders in showcasing local food are headquartered in Virginia: Ukrop's Super Markets, a 30-store chain based in Richmond with outlets in central Virginia, and Food City, a privately held, family-owned supermarket chain based in Abington with nearly 100 stores in southwestern Virginia, northeastern Tennessee, and southeastern Kentucky. In 2000, both firms began purchasing local produce from a southwestern Virginia alliance of organic family farmers that market under the Appalachian Harvest label. About 60 growers in southwestern Virginia and eastern Tennessee now participate in the Appalachian Harvest marketing program.

Other retail customers of the Appalachian Harvest marketing collective include Ingles Supermarkets, a supermarket chain based in Asheville, N.C., and Whole Foods Market stores in the District of Columbia, Pennsylvania, and northern Virginia. Not only have these retailers eagerly embraced the opportunity to purchase local produce, but one firm (Ukrop's) was so interested in establishing a steady supply of locally grown organic produce that it helped the Appalachian Harvest network recruit former tobacco farmers to grow organic produce.[39]

Ukrop's also extended its in-store promotion of local food to include value-added food products manufactured with locally grown ingredients. In 2002, Ukrop's was the firstgrocery to offer in its produce department herb salad dressings from Shenandoah Growers along with national brands.[40] This Harrisonburg, VA, firm is known for its organic culinary herbs. The dressing was displayed in a stand-alone section next to a display of fresh herbs, underscoring the "freshness" of the locally manufactured product, with each bottle of dressing displaying the "Virginia's Finest" State logo.

Freshness is increasingly important to consumers. Growing exposure to farmers markets has led them to expect merchandise of comparable freshness on supermarket shelves, giving local suppliers a competitive advantage.

A firm's size could also determine the types of goods it prefers to buy from smaller suppliers. Research by USDA's Agricultural Marketing Service (AMS) shows independent grocers and regional supermarket chains often prefer to feature branded products from local suppliers as a way to promote their connections to their communities. Offering distinctive local merchandise allows them to differentiate themselves from larger competitors.[38] On the other hand, to the extent that national and large regional supermarket chains are motivated to buy products from smaller scale suppliers at all, they usually do so from a desire to purchase products with specific or exclusive quality attributes, regardless of geographic origin—including items that can be sold under private labels. For smaller producers, private label programs can generate steady business with large customers and eliminate some of the expense of promoting, advertising, and packaging branded merchandise.

PROMOTING LOCAL ORIGIN: CASE STUDY #2: FAMILYFARMED.ORG

In the Chicago area, sixteen retail grocery firms, comprising a mixture of independently owned grocery stores, specialty food retailers, and regional supermarket chains, recently agreed to begin carrying FamilyFarmed.org food. This label indicates the food is grown in Midwestern States by family owned and operated farms that either are certified organic or employ similar sustainable agricultural practices. A nonprofit organization known as Sustain coordinates and recruits participants for FamilyFarmed.org. Participation allows members to:

- Post their farm's story and photo on the FamilyFarmed.org Web site <http:// familyfarmed.org>.
- Participate in an annual exposition where they can market their food directly to household consumers and commercial buyers.
- Use FamilyFarmed.org labels in their product merchandising and advertising. The label effectively communicates their farms as small, locally operated, family businesses that use sustainable production methods. Members can purchase stickers featuring the FamilyFarmed.

org label from Sustain at prices ranging from $25–$50 per 1,000, depending on volume.

Sustain also offers its members customized design services for boxes, bags, and other product packaging on a fee-for-service basis. The designs showcase a farm's personal story on consumer-ready packaging units. Figure 13 shows a custom-designed 5-pound bag for organic potatoes by Igl Farms of Antigo, WI. The FamilyFarmed.org label appears in the lower left-hand corner of the front of the bag. The back of the bag (not shown) features a photograph of the Igl family, a quote from the growers about the mission of their farm, a potato recipe, a Wisconsin State logo, a USDA organic seal, information about the farm's organic certifying agent, and basic nutritional information.

Figure 13. Promotional Materials from FamilyFarmed.org

Retail buyers not only recognize the appeal of locally grown food to their increasingly discriminating clientele, they also realize that the growing popularity of farmers markets raises expectations of freshness in supermarkets. This gives a substantial competitive advantage to local producers that can deliver high-quality seasonal produce. Richard Draeger, vice president of Draeger's Supermarkets, a three-store supermarket chain based in Menlo Park, CA, recently observed to a Supermarket News reporter that "regular customers expect nothing less than ripe, ready-to-eat produce" because they've shopped at farmers markets.[41]

Meeting Retail Demand for Locally Grown Food: A Case Study Of Big River Foods

Big River Foods is the newest program of the Minnesota Food Association, a nonprofit organization based near Stillwater, MN, whose mission is to build a more sustainable food system. Big River Foods is closely linked with the Minnesota Food Association's New Immigrant Agriculture Project (NIAP), established in 1999 to help immigrant and other limited-resource farmers learn sustainable agriculture methods and farm business management. Since its inception, more than 250 immigrants (primarily Southeast Asian (ONEST_ Hmong and Latinos) have benefited from these training sessions. Big River Foods (BRF) became a natural outgrowth of NAIP as a means of meeting the growing marketing needs of a number of the more commercially oriented immigrant farmers, many of whom were looking for marketing outlets beyond farmers markets. BRF is set up as a "training distribution company" that combines brokering functions and transportation logistics with on-farm production and post-harvest handling training.

Big River Farms

Fresh. Local. Honest.

BRF is well positioned to tap into the growing demand by supermarkets and other retail outlets to provide their customers with locally grown and sustainably produced fresh fruits and vegetables. In only its firstyear of operations, BRF has successfully brokered deals and distributed to one regional and two local supermarket chains operating in Minneapolis/St. Paul and the surrounding area. BRF sells several tomato varieties (grape, slicer, heirloom) to Lunds and Byerly's (a high-end supermarket with 21 locations); Kowalski's Market (a supermarket chain with eight locations and a focus on organic and natural foods); and Cub Foods (a regional supermarket chain with 84 locations in the Midwest). It also sells heirloom and grape tomatoes to the Wedge in Minneapolis, one of the largest food co-ops in the country, and green peppers to the restaurant chain Chipotle.

BRF is tapping into only a fraction of this retail demand, currently working with five family farm operations, most of which are growing on less than 2 acres of land. BRF entered the distribution business knowing that demand for locally grown food would be higher than its production capabilities. Nonetheless, it has had a successful start by sticking to some core marketing principles:

- Produce for the market rather than trying to market what you produce. BRF reached out to numerous retail buyers to find out what products were in highest demand. By matching its own production capacity to retail buyer needs, BRF saw a lucrative market for selling tomatoes and green peppers.
- Develop a brand identity that refl ects the core values of the company and appeals to what consumers want in a product. The BRF logo states simply that its products are "Fresh. Local. Honest."
- Live by the standards espoused in your brand. For BRF, this means ensuring its farmers are certified to use good agricultural practices (GAPs), carry liability insurance, and grow their produce using ecologically sound practices.
- Ensure credibility with retail buyers by knowing everything about the products you are selling. BRF decided to limit sales to tomatoes and green peppers. The BRF team knew every aspect of production and post-harvest physiology and handling for these two products. When speaking with retail buyers, BRF could confidently answer questions concerning production capacity, handling, storage, and delivery mechanisms.

- Start small and within your present capacity. With limited cleaning facilities and cold-storage capacity, tomatoes and green peppers were well-suited for BRF. Cleaning, grading, and packaging could be done on-farm, and both products could be stored in cold storage units of similar temperature. The farmers were also comforTable growing these two products at a level that met the quality standards expected from the retail buyers.
- Communication, communication, communication. Cultivating a strong relationship with a retail buyer requires constant communication. This starts by finding out how the buyer prefers to communicate, whether this is by fax, phone, or email. It then means regularly communicating with the retail buyer, from assuring the buyer that product quality and handling standards are met to updating the buyer on orders and delivery dates.

Further information on Big River Foods can be found at the Minnesota Food Association's Web site, http:// www.mnfoodassociation.org

Private Label Products

Private label or store brand programs represent another potential market opportunity for smaller food producers and processors. Such programs appeal to retail buyers who seek food products with specific—and sometimes exclusive—quality attributes. Although in the past private label products had a reputation for inferiority, a 2005 report by Citigroup/Smith Barney reported that sales of private label products in the United States increased 48 percent between 1997 and 2003, compared with a 22-percent increase for branded items and 25 percent for other packaged goods.[42] The retail industry has enthusiastically adopted private label programs in recent years, with the result that, by 2007, one of every five items sold in supermarkets, drug stores, and mass merchandise stores was a private label product, accounting for more than $65 billion in sales per year.[43]

According to the market research firm Mintel, in recent years sales of private label food have outpaced other food products. Sales of the 20 best-selling categories of private label food—items such as cheese, frozen vegeTables, and cookies—rose 23 percent between 1999 and 2004, while overall sales of the same product categories rose only 14 percent.[44] Consumers have generally embraced private label food products—seven out of ten consumers recently told the market research firm Packaged Facts that private label foods are as good or better than national brands.[45]

Private label products are becoming increasingly attractive to supermarket buyers for several reasons. Consumers are shopping at an ever-broader array of food stores, and are able to buy national brand products at a growing number of firms. Therefore, traditional full-service supermarkets can no longer attract customers by featuring a wide selection of national brands and depending on brand loyalty to drive sales. To remain successful, they need to differentiate themselves from other retail outlets. Private label products fit into this strategy by enabling supermarkets to offer exclusive food items that aren't available to their competitors and also directly promote their company's reputation.

Private label products also offer financial and quality control advantages to both retailers and consumers. In the past, private label and store brand programs enabled retailers to pass along savings to their shoppers by eliminating some expenses and frills associated with branded products from major food manufacturers, such as manufacturers' advertising costs or their elaborate packaging. In today's highly competitive retail environment, where big-box and mass-merchandise stores continue to chip away at supermarkets' share of the food market, the lower prices of private label products attract supermarket buyers.

In an era distinguished by consumer attention to food quality, private label products enable retailers to exert greater control over food production and processing methods, and to use this control as a marketing advantage. As one managing consultant put it, "companies [these days] view risk to their brands as their single biggest business hazard."[46] Business alliances that permit retailers to inform consumers about the origin of their food and assure its quality, and to act quickly in the event of a recall, bestow substantial marketing advantages as well as the opportunity to educate customers about the distinctive qualities of their store brand products. Smaller scale food producers and processors are in a position, both geographically and operationally, to develop exclusive relationships with retailers that value—and are willing to pay for—products with assured quality characteristics that can quickly be traced back to their source.

CASE STUDY OF A PRIVATE LABEL: DOROTHY LANE

A recent interview in Supermarket News explored a successful private label marketing campaign for meat that relies upon a direct relationship between a retail grocer and small-scale suppliers of natural meats.[47] As part of its private label program, Dorothy Lane Market, a three-store specialty grocer based in Dayton, OH, offers consumers a choice of several free-range poultry, natural beef, and natural/no-nitrate pork products. The specifications incorporate attributes related to environmental sustainability and humane practices. For example, Dorothy Lane Market's private label free-range turkeys are guaranteed to be:

- Raised "free-range" (meaning, according to the market's definition, the birds are free to range outside, to scratch the ground, and forage on plants).
- Raised without antibiotics.
- Raised without hormones or growth stimulants.
- Vegetarian-fed—with no animal by-products in the feed.
- Locally grown and raised on a small family farm (Bowman & Landes in New Carlisle, OH).

In a February 2005 interview with Supermarket News, Jack Gridley, the director of meat and seafood purchasing for Dorothy Lane Market, said his firm buys beef only from one source: Coleman Natural Beef. All beef entering his company's private label program originates from Coleman in order to ensure traceability of the food products from "farm to plate." Market officials estimate they can identify the farm source for 93 percent of their meat transactions, so if there is a recall, they can respond immediately.

While Dorothy Lane Market's private label program may include unusually stringent quality, animal husbandry, and social welfare standards, a growing number of retailers have begun to incorporate some type of certification or natural food labeling into their private label programs, as such programs have proven to be winning marketing strategies for some early market entrants. Whole Foods Market, the prominent retailer of natural foods, led the charge in developing organic and natural store brand programs back in 1997 and 1998 by launching its 365 Organic and Whole Kids lines of products.[48] By early 2005, these store brands represented about 15 percent of the company's revenue.[49] Several conventional supermarket chains have followed Whole Foods Market's lead by launching their own store-brand organic programs over the past few years, including Kroger Company (Naturally Preferred), Shaw's Supermarkets (Wild Harvest), Giant Food/Stop and Shop (Nature's Promise), Safeway (O Organics), and Wegmans Food Markets (Wegmans Organic).

Retail Concentration

The number of supermarket chains operating in the United States has dwindled in recent years as many of the country's largest supermarket firms have increased their scale of operation and sought to improve profitability by acquiring assets from less successful competitors. Today the U.S. grocery sector is increasingly controlled by fewer and fewer grocery firms (figure 14). Whereas in 1998 the top 20 grocery firms in the United States accounted for slightly more than half of total sales, by 2004 they accounted for more than two-thirds. During the same period, the top four grocery firms increased their market share from 30 percent to almost 40 percent of total grocery sales volume.

Two trends appear to be responsible for the intensified concentration of the grocery sector over the past couple of decades—the growing range of marketing outlets that sell food, which has undermined the longstanding role of supermarkets as a primary source of food, and the declining importance of at-home food preparation, which has boosted sales of ready-to-eat foods (often from foodservice outlets) at the expense of other food purchases.

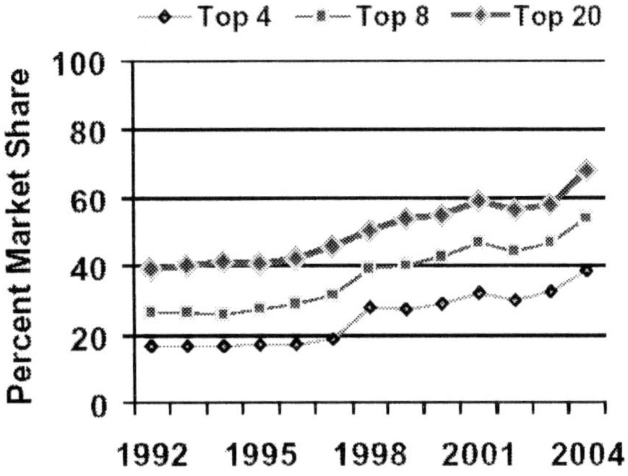

Source. USDA/ERS

Figure 14. Market share of top supermarkets, 1992-2004

Channel Blurring

The growing influence of discount mass merchandise stores in the retail sector has transformed the way people shop for food. In 1990, supercenters—combination retail stores that blend an economy supermarket with a discount department store—sold just under 2 percent of all food sold in the United States. By 2005, their share of the U.S. food market had grown to nearly 14 percent.[50] Their ability to purchase food products in large volume and use their buying power to offer unusually low prices to consumers appears to have earned them a lasting niche in the retail grocery marketplace.

The Squeeze on Grocers

According to recent analysis by ERS, the market for groceries is essentially saturated. Despite steady increases in the U.S. population, the growth potential of retail grocers in the United States remains limited by the declining share of consumer income spent on food.[51]

Further exacerbating the financial pressure on supermarkets has been the steady increase in the share of food dollars on food eaten away from home. Between 1966 and 2006, the share of food dollars spent on food consumed outside the home climbed from 31 percent to nearly 49 percent.[52] For more information about food eaten at home versus away from home see figure 1 on page 7.

In this increasingly competitive marketing environment, supermarkets are looking for ways to improve their profit margins. Supermarkets at all scales of operation are working with their vendors to integrate supply chain management systems and ensure end-to-end quality control. These systems use information technology to streamline the distribution process, and include special arrangements with vendors to increase supply chain efficiency.

Integrated data systems have already led to savings among the early adopters. With the costs of technology continuing to fall, supermarkets will continue to adopt increasingly complex technology to support their distribution operations, especially as the potential for long-term cost savings becomes widely recognized. For more detailed information about technology in the food supply chain, see our companion publication, *Supply Chain Basics: Technology: How Much—How Soon*, available at http://www.ams.usda.gov/MarketingServicesPublications.

How Does Retailing Efficiency Affect Food Suppliers and Processors?

The use of information technologies, from barcodes to RFID and data synchronization, has increased the amount and efficiency of the information supermarkets share with their suppliers—and that they expect in return. Results from a 2000 food industry survey show that more than 80 percent of surveyed retailers demand more services from their suppliers than they did in 1993.[53] The top three services reported by retailers were the provision of private label products, category management services, and electronic data interchange capability.[54] Consequently, to be taken seriously as suppliers in today's retail marketplace, food producers and manufacturers must develop the technological capability to exchange product information electronically.

Beyond anticipating consumer demand and streamlining logistics with information technology, retailers are exercising leverage to negotiate purchases of food products customized to their procurement and marketing needs. To ensure that suppliers meet their standards, some retail firms are initiating Vendor Certification Programs (VCP) that require certified vendors to guarantee specific safety, handling, traceability, packaging, and quality control practices.

To ensure that inventory levels are finely coordinated with demand, supermarkets have made arrangements with some vendors to ship directly to stores instead of to a centralized distribution warehouse. (This movement is discussed in detail in the section "Direct Store Delivery.") The move to customized product offerings for each grocery store opens new marketing opportunities for small- and medium-sized food producers and processors that might not produce enough to supply the large number of retail stores served by a distribution center, but that do have the capacity to supply products to a selected number of individual store outlets.

Advantages of Supply Chain Coordination

The retail food industry is forming tightly aligned supply chains to gain two key objectives—efficiency and improved customer response.

Efficiency

Supply chain efficiencies are gained by effectively sharing information between and within firms. When information is shared among a firm's departments rather than each department operating in isolation, the firm becomes more efficient. A lack of communication among departments slows a firm's response time, possibly marketing products consumers don't want. Departmental integration focuses all departments on the same goal and makes the firm more responsive to customer demand. Efficient internal communications are essential to effective external communication with suppliers.

Data sharing between retailers and their suppliers is not new, but it has become more efficient and richer in detail. The advent of inexpensive computer equipment and software allows increased collection and tracking of many kinds of product data all along the supply chain.

The streamlining of data exchange with suppliers helps supermarkets match their product inventory to consumer needs on a store-by-store basis. At the same time, suppliers benefit from the information-sharing arrangements by learning consumer needs and preferences faster and with greater accuracy. As the cost of technology declines and its capability expands, the trend toward close coordination between retailers and suppliers can be expected to accelerate until it becomes the new standard for partnerships between supermarkets and their suppliers in firms of all sizes.

Improved Customer Response

Firms today aspire to become demand-driven supply networks, with the agility and efficiency to succeed in a rapidly changing global economy. This business model is driven by capturing information about consumer demand, not by forecasts or plans. Suppliers with real-time data collected at point of sale can quickly modify their products, quantities, and package sizes based on customer preferences. For more detailed information about electronic exchange of information along the supply chain, see our companion publication, *Supply Chain Basics: Technology: How Much—How Soon.*

Traits of Top Grocery Firms

Top grocery firms share a commitment to communication and integration with their suppliers through adoption of information technology, which can lead to cost reductions for business operations as well as increased sales. Streamlined data and information exchange with their suppliers help retail firms customize their product assortments to consumer needs on a store-by-store basis. On the other hand, producers benefit directly from information-sharing arrangements with retailers by developing a more precise understanding of consumer needs and preferences quickly. As the cost of technology declines and its capability expands, the trend toward close coordination between retailers and suppliers can be expected to accelerate and become the new standard for business partnerships between retailers and vendors, regardless of firm size. Consequently, small and mid-size food producers interested in supplying the retail grocery market will need to employ such technologies as barcodes or RFID tags that enable them to share product data throughout the production, handling, and distribution processes.

Global Data Synchronization

In the modern world of global sales, it has become imperative for businesses in different countries to coordinate their supply chain data. This coordination is called *Global Data Synchronization* (GDS). The goal of GDS is to create a universal product code that can be understood by sellers and buyers everywhere. With such a code, stores and their suppliers can communicate clearly, unambiguously, and—most important—accurately about the goods they are buying and selling.

How Does Data Synchronization Work?

Data is exchanged between retailer and supplier through data pools, which provide transparent gateways through which suppliers and retailers exchange close-to-real-time information about products, packaging, and logistics. Before data are made available for access, they are checked for accuracy and configured into standard formats. The data pool defines each data element to ensure that buyer and seller are "speaking" the same language. This standardized data format provides accuracy and allows buyer and seller to know exactly where to locate information.

GDS began with a program called Efficient Consumer Response (ECR). ECR programs, introduced in the 1990s, were designed to allow a seamless fl ow of product from producer to consumer, minimizing inventory holding costs while reducing the number of out-of-stock items and maintaining high levels of customer satisfaction. ECR ties grocery manufacturers, wholesalers, and retailers together with information systems. The four basic strategies of ECR are:

- Efficient store assortment.
- Efficient replenishment (Just-in-time inventory minimizes the amount of inventory held by distribution centers and store back rooms).
- Efficient promotion.
- Efficient product introduction.

Prior to ECR, supermarkets and food manufacturers had focused primarily on reductions in logistical expenses.

GDS is an extension of ECR. Like ECR, it is designed to reduce excess inventory but, because it reduces the incidence of data errors and allows data to be shared more efficiently, it saves time and money throughout the supply chain process. The advent of computerization, the Internet, and the declining cost of high-powered computer equipment and software have made this type of constant communication and intensive inventory management increasingly possible and affordable.

The Global Data Synchronization Network

In 2003, FMI and the Grocery Manufacturers Association (GMA) sponsored six case studies measuring the impact of data synchronization on grocery firms and food manufacturers between November 2002 and January 2003. Three manufacturers and three retailers were selected, ranging from relatively small regional firms to large conglomerates: Bi-Lo (a discount retail grocery subsidiary of Ahold USA) Kraft Foods, Nestle, Purina Pet Care, Proctor and Gamble, Shaw's Supermarkets, and Wegmans Food Markets. Each of the firms in the case study had implemented data synchronization. The resulting report is titled *Data Synchronization Proof of Concept: Case Studies from Leading Manufacturers*.[55]

The findings of the study were promising and resulted in the launching of the Global Data Synchronization Network (GDSN), an electronic system that exchanges data between suppliers and retailers using a standardized format. GDSN is the electronic network that makes GDS practical, allowing suppliers and retailers to communicate product information accurately, quickly, and efficiently. Early adopters of GDSN have experienced significant time and cost savings and want their suppliers to participate in data synchronization.

As the technology spreads to other supermarkets, more food buyers will require their vendors to participate in GDSN. Since mid-2003, food industry participation in GDSN has risen from 100 companies to nearly 10,000, and the number of registered food items grew from 60,000 to more than 580,000 in 2005.[56] GDSN is increasingly a requirement for food suppliers.

Overview of GDSN

GDSN is a network of data pools with a global registry for sharing information about products and companies. It was conceived, and is supported by, the European Article Number (EAN) Association, the Uniform Code Council (UCC), and leading companies and industry groups worldwide. It enables trading partners in the supply chain to exchange data that is accurate, up-to-date, and compliant with universally supported EAN and UCC system standards[57]. GDSN is based on a centralized global registry that connects to data pools around the world, enabling standardized data to be synchronized on a near real-time basis.

GDSN was developed in response to the high costs associated with inaccurate data. To ensure GDSN meets the business needs of the user community, EAN and UCC established an oversight committee of 17 senior executives from manufacturing, retailing, and EAN member organizations to govern GDSN, promote its global adoption, and address strategic issues related to its rollout.[58] More information about GDSN can be found in our companion publication, *Supply Chain Basics: Technology: How Much—How Soon.*

Initially, it was thought the benefits of GDS would come from improved standards, accuracy, and synchronization. However, improvements have surpassed these basic areas and include such improvements in vital processes as demand forecasting, procurement, replenishment, and logistics. Researchers believe this is just the tip of the iceberg, and still more improvements in RFID, Electronic Product Code (EPC), and price synchronization can be expected as GDS becomes the standard for the retail grocery sector.

What are the Advantages of GDS to Retailers?

- *Reduced inbound and outbound freight costs.* GDS improves the accuracy of product weights and measures information resulting in truckloads more completely filled but not exceeding weight limits. The improved efficiency results in lower transportation costs.
- *Improved productivity within distribution networks.* Labor costs are reduced due to fewer work hours needed to correct and align product information between retailers and suppliers.
- *Improved DSD receiving capabilities.* More accurate data means that more items scan correctly at the back door and fewer labor hours are required to correct errors.
- *Improved productivity within order and item administration.* Correct item information streamlines administrative activities, resulting in improved response time of continual replenishment programs, less time for order inspection, and reduction in labor hours required to process new items.
- *Reduced need to adjust and reconciliate invoice and coupon discrepancies*, which lowers labor costs.
- *Reduced time from item entry to retail shelf.* (Speed to market and customer service).
- *Reduced time spent addressing item data issues.*

What are the Advantages of GDS to Suppliers?

- Reduced inbound and outbound freight costs.
- Improved productivity within distribution networks.
- Improved productivity within order and item administration.
- Reduced time from item entry to retail shelf.
- Reduced time addressing item data issues.
- Makes your company more appealing to retailers who have or will soon adopt GDS.

How does GDS Better Utilize Current Technology?

Prior to GDS, suppliers and retailers created, stored, and used electronic data exclusively for their internal operations. Retailers' and suppliers' product data had no standard format, data fields varied from firm to firm, and the information collected was not generally shared among companies.

Data synchronization organizes into a standard format product data that has been captured electronically by scanning barcodes, uniform product codes (UPC), or RFID tags. The stored information can be accessed on the network by both supplier and retailer. Because product information is synchronized, both have the same information in a standardized format, making it easy to find and understand product information. For example, when a supplier changes any catalogued characteristic of a product on GDSN, the new product characteristic is shared with the retailer almost immediately, eliminating errors due to incorrect information. Retailers know exactly what they are getting and they get the products they ordered. The synchronized data makes the supply chain transparent to buyers and sellers, reducing confusion and costly errors.

When a retailer receives a shipment utilizing GDSN, these are the benefits when an electronic invoice is scanned:

- The retailer saves the time it takes to record the shipment manually;
- Errors are avoided that could occur by manually recording the contents of the shipment;
- The shipment can be compared electronically with the original order.

When the retailer acknowledges receipt to GDSN, the supplier sees the receipt and the date and time received.

Unexpected Benefits of Data Synchronization

In the spring of 2006, sponsored by GMA, FMI, Wegmans Food Markets, and 1SYNC, a GDSN data pool, a follow-up to the 2003 study investigated the benefits of GDSN. The resulting report, *Synchronization—The Next Generation of Business Partnering,*[59] not only confirmed the findings of the 2003 initial study but surpassed some original projections. It found that:

- Using accurate data to drive business processes develops value for all industry players.
- Data synchronization is not only technology—it is people and processes working together to form a highly integrated and collaborative value chain.
- Manufacturers and retailers that lead the way are realizing significant benefits.

Table 4. Benefits of Data Synchronization

Performance Metrics (Process Area)	Source of Benefits[a] % Improvement over Current State	
	Retailer[b]	**Manufacturer[c]**
Reduced Inbound and Outbound Freight Costs (Transportation)	6.5%	1–8%[d]
Improved Productivity Within Distribution Network (Distribution)	1%	64%[e]
Improved DSD Receiving Capabilities (Store Operations)	9%	Not Applicable
Improved Productivity Within Order and Item Administration (Order Administration)	50%	2–67%
Reduced Out-Of Period Adjustments and Reconciliation of Invoice and Coupon Discrepancies (Accounting Administration)	8%	Not Applicable
Reduced Time from Item Entry to Shelf (Speed to Market)	23%	67%
Reduced Time from Item Entry to Shelf (Customer Service)	40%	Not Applicable
Reduced Time Spent Addressing Item Data Issues (Sales)	Not Available	25-55%

Source. *Synchronization—The Next Generation of Business Partnering*, Accenture, 2006.

a. Where actuals were not available, estimates were provided

b. Percentages were calculated using weighted averages, based on dollar values

c. Represents spot benefits for particular manufacturers; percentages do not reflect savings across all participants

d. Eight percent based on mill transit savings for a single product line at a particular manufacturer. Total mill transit savings by this manufacturer is pending completion of a full product portfolio audit and is expected to be lower.

e. Improvement is in a particular area at manufacturer; does not represent improvement across entire distribution network.

Not only have manufacturers improved productivity, but they have also reduced the time it takes for a new item to reach the retail shelf. Due to greater efficiencies in the distribution network, one surveyed manufacturer was able to condense a spreadsheet-based process used to align product information with its distributors from 5 days to 48 hours.[60] Some significant productivity improvements include reduction in the time needed to add new products; change existing product characteristics such as size, case count, and palletization requirements; reduced instances of coupon rejections at store check-out; and reduced time for a new item to reach the shelf.

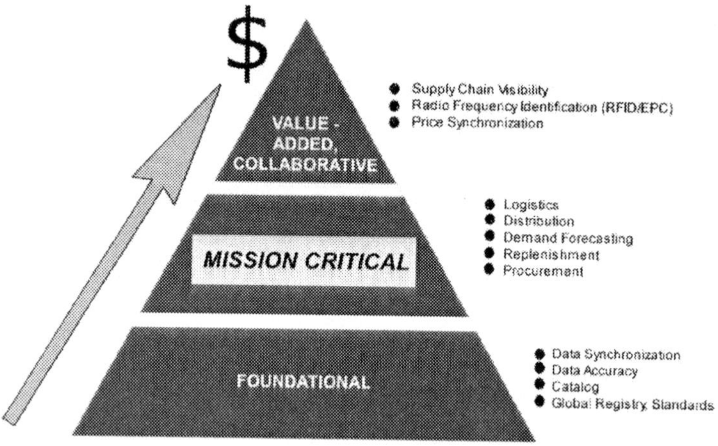

Source. *Synchronization—The Next Generation of Business Partnering*, Accenture, 2006.

Figure 15. E-Collaboration Pyramid

One participant in the GMA/FMI synchronization study introduced 720 new products into its portfolio with significantly less labor than projected, saving 3,480 hours of employee time annually—an estimated savings of $288,000.[61] This sizable addition to its portfolio represented a 50-percent increase in its product line. Prior to the implementation of GDSN, the process of assembling and circulating all relevant data associated with these products to internal units and trading partners would have required hiring an additional two full-time employees to enter the data, print or develop documentation, and send it out.[62]

Initially, most benefits were expected to be derived from improved accuracy, synchronization, and standards for product registration and changes. However, those savings were just the beginning. Benefits also are achieved through enabling employees to concentrate on such "mission critical" tasks as demand forecasting, procurement, replenishment, and logistics, rather than validating item information and doing other administrative paperwork. Ultimately, even greater benefits may emerge as industry participants continue to work together on collaborative processes such as RFID, EPC, and price synchronization.[63]

Figure 15 illustrates improvements in the supply chain brought about by data synchronization. The 2006 Accenture report noted that initial improvements such as data accuracy and common global registry standards are the catalyst for improvements and savings in logistics, distribution, demand forecasting, replenishment, and procurement. These second-tier improvements, in turn, pave the way for enhancing the ability to track and control products through the supply chain, adoption and enhancement of RFID technology, and the development and use of price synchronization.[64]

Even in the initial stages of adoption, GDSN is already providing significant benefits to the food industry. For example, prior to the implementation of GDSN, one of the food manufacturers featured in the study had been fined $300,000 to $500,000 a year by retailers for inaccurate product information. After the adoption of GDSN, the fines assessed on the same manufacturer for inaccurate information were close to zero. Another food manufacturer improved the time needed to move new items from 4 to 8 weeks to only 2 weeks. Yet another study participant saved $2.2 million in annual transportation costs by correcting a weight error on a single item.[65]

Wegmans Food Market Experience with Data Synchronization

As of the summer of 2006, 95 percent of Wegmans Food Markets' vendors were participating in its data synchronization program, 92 percent had actually traded at least one item through their synchronization network, and 66 percent were completely synchronized. The value of benefits to Wegmans Food Markets from more accurate data is estimated at $3.5 million in annual transportation costs. In addition, new items arrive on retail shelves about 7 days quicker, labor and inventory carrying costs are reduced by about $1 million annually, and coupon scanning rejections are cut by about 40 percent.[66] This translates to cost reductions of $500,000 for every additional $1 billion in sales, as well as increased simplicity throughout the supply chain process.

Overall, food manufacturers participating in Wegmans Food Markets' data synchronization program are projected to reap $1 million in savings for every billion in sales. This increase in earnings represents estimated savings from reducing bloated inventory levels and stocking problems due to inaccurate data.[67]

Direct Store Delivery

While supermarkets generally prefer "just-in-time" inventory practices for budgetary and logistical reasons, maintaining low inventory levels throughout the distribution system has the potential to increase out-of-stock levels at individual retail stores, which can be a financially risky undertaking. Consumers' disappointment when they don't find the items they want on retail shelves is estimated to cost manufacturers $28 million and retailers $32 million for every billion dollars of sales.[68] According to a recent Franklin Covey study of the U.S. retail industry, called *Mapping Management Practices that Drive Great Performance*, the typical out-of-stock level for retailers is about 8.5 percent.

There are many causes for out-of-stock items: inaccurate forecast of consumer demand, shipment by the supplier of the wrong package size or flavor, unfit products due to product damage during shipping, expired product shelf life, or the seller's inability to track inventory accurately in its stores or distribution centers. Given the frequency of out-of-stock problems, and the huge financial burden that large out-of-stock levels can create for supermarkets, suppliers able to help supermarkets avoid out-of-stock items are likely to attract retail food buyers. Consequently, some enterprising supermarket chains are beginning to explore using *Direct Store Delivery* (DSD), an arrangement in which suppliers deliver directly to stores instead of to a distributor. DSD helps supermarkets overcome out-of-stock problems while keeping minimum inventory at their distribution centers.

The move to greater acceptance of DSD among retail chains may provide expanded market access to smaller scale local food producers and processors with the capacity to supply food items appealing to specific segments of a firm's customer base, but may not have the capacity to supply products to every store supplied by the chain's centralized distribution center. Recent market research by AMS suggests food items that perform extremely well in certain markets may not be as profiTable for a broader sales territory.[69] Firms that are open to receiving food items via DSD and tracking product performance on a store-by-store basis may be more accommodating of smaller scale local producers than firms more heavily dependent on centralized procurement mechanisms. Indeed, a straw poll taken during a recent webcast sponsored by Supermarket News discovered that food retailers found it easier to collaborate with suppliers at the store level with DSD than with warehouse deliveries. Sixty percent believed DSD provided easier collaboration, 20 percent found warehouse deliveries better, and 18 percent found no difference.[70]

Producers and processors of locally grown and raised food may find it easier at firstto approach retail firms that showcase foods from the nearby region. For example, Whole Foods Market, as part of its marketing philosophy, is strongly committed to purchasing locally grown produce. Not only do the firm's regional buyers purchase locally grown products, but the firm also allows individual stores to purchase selected merchandise directly from local producers via DSD, tailoring product offerings in its store to the preferences of its particular market area. In 2006, regional buyers from Whole Foods Market were explicitly directed to focus more attention on sourcing local products for their stores, and subsequently obtained merchandise from more than 2,400 independent farms.[71] Purchases from individual and family farmers (including marketing alliances of family farms pooling production under a common brand name) accounted for 78 percent of Whole Foods Market's total fresh produce purchases in 2006.[72]

In addition to purchasing products from local growers, Whole Foods Market established a $10 million promotional program for local agriculture, which offers low-interest, long-term loans to small agricultural entrepreneurs.[73] The loan program targets local producers of grass-fed beef, goat milk dairies, pasture-based eggs, and farmers with animal-compassionate policies. A portion of the loans will be available to vegetable farmers as well. Each year, Whole Foods Market makes an additional $10 million available for these loans. As the loans are repaid, the funds are recycled for additional loans to small producers.

A Case Study of Market Access Options for Small Farmers: Whole Foods Market, Orange County, CA

During a March 2007 seminar, Jeff Biddle, Whole Foods Market's produce coordinator for the Pacific region, and David Schaner, a farmer/roadside stand manager who sells fresh produce to a single Whole Foods Market store near his farm in Placentia, CA, shared tips on how small farmers and food suppliers can successfully enter the retail market.[74] Schaner is among more than a dozen Southern California farmers tapped by Whole Foods Market to supply produce. Since entering into a direct marketing relationship with a single Whole Foods Market store in Tustin,

CA, in 2005, Schaner has boosted his annual revenue about 20 percent. Moreover, the in-store publicity received from point-of-sale material attracted new customers to his roadside stand.[75]

Biddle and Schaner point out that the most important firststep for producers is to recognize that the typical retail buyer is an inherently conservative decision-maker, and can't afford to take chances. When one is moving large quantities of product, each additional penny or nickel per pound adds up to a lot of potential loss. In addition, the daily life of a retail produce buyer is full of anxiety (e.g., weather problems, delays in transportation) and it's always a struggle to figure how best to manage inventory—whether to buy more to avoid running out, or to "short" the stores to save money. No matter what the decision, someone will complain when planned targets for profit margin or shrinkage reduction are not met. On the other hand, there is constant pressure to try new products.

So how can a producer expect to stand out and get his or her foot in the door? Biddle and Schaner recommend the following steps:

- Approach store-level department managers as an initial step in gaining market access, as this tends to be more effective than contacting buyers at a centralized distribution center (who are constantly barraged by prospective vendors.) Even at firms that generally embrace DSD, such as Whole Foods Market, Biddle notes some store managers are more open than others to DSD. Nevertheless, he encourages producers to keep "knocking on doors" even if they do not succeed with the firststore contacted. Once a producer establishes a track record with a buyer for quality and reliability, buyers are usually willing to do what it takes to maintain a successful business relationship.
- Send food! Samples of food accompanied by a subsequent phone call are probably the most effective way to grab a buyer's

attention, and e-mail solicitations are usually ignored. (Schaner developed his current retail business opportunity through a Whole Foods Market representative who had patronized his roadside stand and enjoyed his product.)

- Have information at your fingertips about
- Pricing of your product
- Pack size and packing material
- How the product is cooled
- Product quality standards
- Special handling needs, if any
- Delivery capabilities
- HACCP certification
- Liability insurance coverage (often mandatory)

- Make a point of mentioning organic certification or any humane or environmentally sensitive agricultural practices you follow. Biddle remarked that one produce supplier he worked with near San Diego boosted his sales by publicizing that he deliberately used production methods and barriers that would not harm the rabbits that visited his fields!
- Offer product demonstrations at stores and interact with customers.
- Be willing to open your farm to tours by retail buyers; buyers often become supporters after visiting a farm.
- Don't be embarrassed about charging what you need to make a living. You shouldn't lower the cost of your product just to make a firstsale or enter the retail marketplace, because raising your price after giving an initial lowball figure can alienate a customer.
- Indicate your willingness to provide the retailer special services, such as deliveries, in-store product demonstrations, price lookup stickers on bulk produce, and value-added processing services.

DSD and the Red Tomato Distribution Model: Strategies for Managing Growth

Growers may find DSD deliveries a marketing advantage when shipping to a small number of store locations. Growers with insufficient volume to supply a chain store distribution center may find that they can fill the needs of two to three stores through direct delivery. As growers scale up the number of stores they service with DSD deliveries they may find that DSD is no longer as feasible.

This is the situation in which Red Tomato, a Massachusetts-based nonprofit organization, found itself when it arranged direct deliveries of locally grown produce to 40 retail stores three times a week in the Boston and Philadelphia areas. Among the services Red Tomato provides to its partner growers are product consolidation, repackaging, and transportation to destination markets. At firstRed Tomato viewed DSD as an excellent way for farmers to improve producer access to retail markets and get their foot in the door. However, as the demand for locally grown products increased and retailers wanted to expand delivery beyond the pilot stores, it became too expensive to continue.

In order to continue DSD, Red Tomato addressed the distribution needs of its growers by partnering with transportation service providers, and with retail firms making deliveries to local stores, independent truckers, and farmers who had space after their rural deliveries. Red Tomato found that piggybacking on existing transportation networks helped both parties. Shipping products through their business partners reduced Red Tomato's shipping costs and also improved the operational efficiency and revenues of the shipper by filling trailers that previously were partially loaded or that otherwise would return empty.

To determine if DSD is appropriate for your firm, consider your scale of operation—including labor requirements, transportation costs, and time requirements.

Preferred Vendor and Vendor Certification Programs

As food safety assurances become more influential on purchasing decisions, supermarkets are seeking out food suppliers with track records of proper postharvest techniques, good quality control throughout the distribution process, and recordkeeping procedures that enable supermarkets to track food products to the source efficiently. To accomplish this and reduce the risk of inadequate service from suppliers, many retailers created preferred vendor programs, which restrict vendor selection to suppliers that meet prescribed quality, quantity, packaging, and delivery requirements. Suppliers who participate in preferred vendor programs must adhere to strict requirements in order to maintain preferred vendor status.

Some retail firms prefer to work only with vendors that have enrolled in a formal Vendor Certification Program (VCP). A VCP allows retail buyers to verify their food suppliers comply with accepTable food handling, product quality, and food safety standards.[76] Usually carried out by a third-party auditor, a good VCP starts in the field with a review of the harvest, handling, packing, and shipment practices followed by a grower for a particular food product, and tracks the handling and distribution process of the food product until it reaches its final destination. Some growers develop their own certification programs to monitor performance, which also becomes a marketing tool for prospective customers. These programs are usually verified by third-party inspectors.

The basis of many certification programs is a Hazard Analysis and Critical Control Point (HACCP) plan. HACCP is widely accepted as a means to prevent contamination at vulnerable points in the handling and distribution process where the possibility is greatest for accidental or intentional contamination.

Growers' documentation of their production schedules, handling practices, and adherence to food safety protocols is an essential component of meeting retail procurement requirements for traceback capabilities and preparing for possible recalls. These may include product production dates, use-by codes, daily or hourly production run codes, quantity-produced data, and metal detector verification reports. Carton information should also include the processor name, hour of production, expiration date, and plant ID number to help retail buyers trace products to their places of origin.

A Case Study of Vendor/Buyer Relationships: The Appalachian Sustainable Development Network

The Appalachian Sustainable Development (ASD) Appalachian Harvest brand of organic fruit and vegeTables is supplied by 60 farmers in southwestern Virginia and northeastern Tennessee, many of whom had converted tobacco acreage to organic fruit and vegeTable production. While this network does not claim to have a VCP, it pays close attention to the quality of products and the requirements of grocery retailer clients. Anthony Flaccavento, Executive Director of ASD, says, "If you mostly deliver what grocers want most of the time, they will work with you."

ASD sales are growing rapidly, with 2007 sales expected to reach $700,000, and 2008 sales projected at $1 million. Flaccavento noted that the relationship between ASD and its grocery retail clients had become easier over time. He attributes the improved relationship to ASD's proven ability to meet grocers' specific production, packaging, and shipping requirements successfully time and time again. Flaccavento projects consumer demand in 2007 for his network's production to be more than twice the volume of available supply and estimates that 25–50 growers need to be added to the network to meet current consumer demand. More information about ASD can be found at http://www.appsusdev.org/.

Conclusion

With retail establishments of all sizes, from national chains to independents, seeking to carve out a piece of the natural and organic food pie, food manufacturers who can deliver products that meet their standards at reasonable prices will find a ready market. How can this benefit smaller scale food producers and processors? They may find it cost effective to produce packaged food items for a private label program, and let the retailer bear the financial burden of promoting and advertising the merchandise. Moreover, as private label programs for organic and natural foods become more widespread, some retailers, following the lead of such independent operators as Dorothy Lane Market, may try to distinguish themselves from competitors by involving smaller scale, local businesses in their private label programs. The local businesses then could share "personal stories" with consumers through promotional materials and appeal to consumer desires for local origin and authenticity.

The waning importance of national brands and neighborhood full-service supermarkets in the food marketplace and the rise of increasingly fragmented and diverse consumer shopping patterns offer exciting new opportunities for the smaller scale niche producer and manufacturer of food products. The rapid rise in Hispanic, Asian, and African-American consumer purchasing power, coupled with widespread consumer interest in gourmet and specialty food items, potentially benefits the smaller scale food producer and processor that can supply products that appeal to consumer desires for variety, innovation, authenticity, and freshness.

At the same time, health, nutrition, and environmental issues are becoming increasingly important to consumers who seek to purchase goods in accordance with their social values. Food producers and processors who can provide assurances of environmental stewardship and food safety can be expected to prosper. Retailers of all sizes seek suppliers that can deliver reliable volumes of high-quality, nicely presented food products that satisfy the concerns of environmentally and health-conscious customers. If demands for volume, supply availability, and consumer-ready packaging can be met, local smaller scale food producers and processors can play an integral role in today's rapidly diversifying food supply chain. Whether their participation involves producing branded food products for specialty retailers that use local origin as a consumer "hook," or supplying customized/exclusive food products to a retailer's private label program, smaller scale food producers and processors have many options for filling an important niche in the retail food marketplace.

The U.S. grocery sector is a mature market and, as such, its growth rate is limited to the growth in U.S. population. Consequently, one of the new frontiers in the grocery sector for increased sales and profits and reduced expenses is in the arena of data collection, accuracy, and exchange. Improvements to information transmission throughout the supply chain are expected to lower warehousing, distribution, and marketing costs.

Gains in sales are facilitated by more efficient collaborative relationships between suppliers and supermarkets from the adoption of data collection and data synchronization technologies. Leading supermarkets and suppliers electronically capture and analyze near real-time data throughout the supply chain, giving supermarkets the ability to quickly modify product offerings, quantities, package sizes, and other product attributes that consumers prefer. Collaboration between supermarkets and suppliers assists supermarkets' efforts to consistently merchandise a high-quality product mix that accurately refl ects consumer desires at the store level. Food producers and manufacturers with proficiency in electronic data collection, sharing, and synchronization, and that collaborate with buyers, have a marketing edge in the grocery sector of the new millennium. Over time, as the collaboration between supermarkets and suppliers increases and the cost of technology declines, compatible forms of electronic data transmission will become even more critical.

Accurate evaluation of customer demand and precise tracking of inventories ensure that supermarkets have the right product on the shelf. Simultaneously, supermarkets will limit inventory stored in back rooms and distribution centers to precise levels sufficient only to prevent running out of stock, and will collaborate with vendors. Market leaders in the grocery sector are already committed to the Information Age, and require vendors to collect and record information about food products from the time crops were planted or animals raised through final delivery.

Product knowledge alone, however, is not sufficient in today's market. Suppliers must also be sensitive to the market demand for their product and know how the supermarket or consumer intends to use the product. Suppliers that develop this capability will be able to respond more quickly to shifting consumer preferences and retain a marketing advantage in the rapidly changing grocery sector.

End Notes

[1] Derived from presentation delivered at Produce Marketing Association convention, March 2008.

[2] "Break it up: deconsolidation and decentralization will dictate retailing's future," Progressive Grocer, February 15, 2003, p. 50.

[3] "American Grocery Shoppers Seek Quality, Value, and Convenience from Multiple Retail Formats, According to FMI's Trends 2005." FMI.

[4] "Poll Reveals Consumers Are Increasingly Exploring Warehouse Clubs, Supercenters and Other Alternative Grocery Formats, But Supermarkets Remain Preferred Source of Food Purchases," FMI and SupermarketGuru.com, August 13, 2002.

[5] Ibid.

[6] "Food Retailers Offering More Takeout, Self-Scanning, Fuel Pumps, Drive-Thru Pharmacies, and Other Services fo Convenience-Driven Consumers," FMI, May 2, 2004.

[7] AMS. Data current as of September 2007.

[8] *USDA Releases New Farmers Market Statistics*, AMS, Washington, DC, December 5, 2006.

[9] *Robyn Van En Center for CSA resources*, Wilson College. <www.wilson.edu/csacenter>.

[10] U.S. Organic Industry Overview, excerpted from the Organic Trade Association's 2006 Manufacturer Survey. <www.ota.com/pics/documents/ short%20overview %20MMS. pdf>.

[11] Food and Alcoholic Beverages: Total Expenditures, Food CPI, Prices and Expenditures Briefing Room, ERS <www.ers.usda.gov/briefing/cpifoodandexpenditures/Data/>.

[12] *Locally Grown Foods Niche Cooks Up at $5 Billion as America Chows Down on Fresh!*, <www.packagedfacts.com/anout/release.asp?id=918>.

[13] National Grocers Association/SupermarketGuru.com consumer panel survey, Fall 2003, reported in "When shoppers stray," Progressive Grocer, April 1, 2004, p.10.

[14] Based on 2007 U.S. population growth forecasts published in The World Factbook, Central Intelligence Agency. <https://www.cia.gov/library/ publications/the-world-factbook/print/ us. html>.

[15] "Independent and Large Chain Retailers Lead Strong Sales Growth; Energy and Credit Card Costs Top Industry Issue Agenda, According to 'Food Retailing Industry Speaks 2006', FMI , May 7, 2006.

[16] "Grocers' real sales growth at 6-year high in '06", Food Institute Report, January 29, 2007.

[17] "Food CPI, Prices, and Expenditures: Analysis and Forecasts of the CPI for Food," ERS. <http://www.ers.usda.gov/Briefing/CPIFoodAndExpenditures/consumerpriceindex.htm>.

[18] Ibid.

[19] "FMI Speaks—Executive Summary: Measuring a Complex Year," FMI, May 2005.

[20] "Excerpted from speech delivered by Carl Steidtmann, Chief Economist and Director of Consumer Business Research with Deloitte Touche Tohmatsu at the FMI Midwinter Executive Conference, January 2005, reprinted in "Trends Indicated Growing Opportunity for Niche Retailers," Supermarket News, January 31, 2005.

[21] "Wal-Mart dominates," Rod Smith, Feedstuffs, July 11, 2005, p. 7.

[22] Ibid.

[23] Supermarket News, April 5, 2005.

[24] "Win at the grocery game: How to shop smarter, cheaper and faster, Consumer Reports, October 2006. <www.consumerreports.org/cro/money/shopping/where-to- buy/ supermarkets -10-06/ overview/1006_supermarkets_ov1.htm>.

[25] "Smaller Food Retailers More ProfiTable Than Larger Chains Last Year," from on-line edition of Progressive Grocer, March 17, 2005.

[26] "FMI Facts About Store Development 2006: New Supermarkets Deliver a Diverse Mix of Fresh Foods," Food Export, March 2, 2007. <http://www. allbusiness.com/agriculture-forestry-fishing/4437834-1.html>.

[27] Progressive Grocer, March 1, 2005, reporting projections by the market research firm Packaged Facts.

[28] "Organic Sales Continue to Grow at a Steady Pace," Organic Trade Association, May 7, 2006.

[29] "U.S. Organic Industry Overview," Organic Trade Association 2006 Manufacturer Survey. <http://www.ota.com/pics/documents/short%20overview%20MMS.pdf>.

[30] Ibid., pp. 5–6.

[31] Ibid., p. 7.

[32] Organic Food & Beverage Trends 2004: Lifestyle, Language and Category Adoption," The Hartman Group, Summer 2004, pp. 26–27.

[33] "Nearly 20% of Americans Can be Considered 'Gourmet' Consumers, New Report Finds," PR Newswire, August 18, 2005.

[34] Ibid.

[35] Ibid.

[36] Source, figures 7–9: "Fastest-growing consumer markets in 1999-2004," Progressive Grocer, February 15, 2005, citing the Selig Center for Economic Growth, Terry College of Business, University of Georgia.

[37] FreshTrack 2001, Cornell University (Includes only firms with annual sales more than $1.5 billion).

[38] *Enhancing Commercial Food Service Sales by Small Meat Processing Firms*, AMS, March 2004.

[39] "Tobacco Farmers Seek Diversification after Buyout," Associated Press, February 19, 2005.

[40] "Ukrop's Teams Herbs, Dressings in Produce," Supermarket News, June 3, 2002.

[41] "Produce Profitability Depends on Freshness," Lynne Miller, *Supermarket News*, April 18, 2005, p. 50.
[42] *Supermarket News*, April 5, 2005.
[43] Private Label Manufacturers Association <http://plma.com/storeBrands/sbt07.html>.
[44] *Baltimore Sun*, May 4, 2005.
[45] Excerpts from "Private Label Food and Beverage in the U.S.," Packaged Facts, Rockville, MD, February 2007. <http://www.packagedfacts. com/Private-Label- Food-1282369/.
[46] Greg Pellegrino, Global Managing Director with the consulting firm Deloitte Touche Tohmatsu, quoted in "Prospering in the Secure Economy," *Meat and Poultry* magazine, October 2004.
[47] "Organic, Natural Meat Sales Exploding," Bobbie Katz, *Supermarket News*, February 2005.
[48] Timing of private label programs obtained from Whole Foods Market.
[49] "Whole Foods has Another Robust Quarter," Austin *American-Statesman*, February 28, 2005.
[50] "Competitive Edge," Willard Bishop Consulting, Ltd., July 14, 2005, and "The Future of Food Retailing," Willard Bishop Consulting, Ltd., July 2006.
[51] *Competition Alters the U.S. Food Marketing Landscape*, Steve W. Martinez, ERS, November 2003.
[52] Food, CPI, Prices and Expenditures: Foodservice as a Share of Food Expenditures <http://www.ers.usda.gov/Briefing/CPIFoodAndExpenditures/Data/Table12.htm>.
[53] "Understanding Retailer Purchasing Trade Practices in the Produce Industry," Willard Bishop Consulting Web site, June 2000.
[54] National Food and Agricultural Policy Project Policy Briefing Paper, March 2001.
[55] GMA-FMI Trading Partner Alliance: Action Plan <http://www.1sync.org/documents/ GDSN/ GMA-FMI% 20Data%20Sync%20Action%20Plan.pdf>.
[56] Synchronization: *The Next Generation of Business Partnering*, Accenture, 2006. <http://www.accenture.com/Global/Services/By_Industry/Consumer_Goods_and_ Services/R_and_I/SynchronizationPartnering.htm>.
[57] Global Data Synchronization Network FAQs, Drummond Group. <http://www.drumm ondgroup. com/html-v2/gdsn-faq.html>.
[58] Ibid.
[59] Synchronization: The Next Generation of Business Partnering, Accenture, 2006. <http://www.accenture.com/Global/Services/By_Industry/Consumer_Goods_and_ Services/R_and_I/SynchronizationPartnering.htm>
[60] Ibid.
[61] Ibid.
[62] Ibid.
[63] Ibid.
[64] *Synchronization: The Next Generation of Business Partnering*, Accenture, 2006.
[65] Ibid.
[66] Ibid.
[67] "In Sync: Wegmans and P&G Collaborate on Product Data, Grocer and CPG Giant Prevent Discrepancies," *ERI Journal*, July 2006.
[68] Presentation by Jim Flannery, Proctor and Gamble, for the SN Interactive Forum on Maximizing Store Performance via Trading Partners Collaboration, June 28, 2006.
[69] *Enhancing Commercial Food Service Sales by Small Meat Processing Firms*, AMS, March 2004.
[70] *Maximizing Store performance via Trading Partners Collaboration*, Supermarket News webcast, June 2006.
[71] Open letter to Michel Pollan from John Mackey, CEO and co-founder of Whole Foods Inc.
[72] Ibid.
[73] Ibid.
[74] Remarks excerpted from presentation delivered at Southwest Marketing Network Conference, Flagstaff, AZ, March 2007.

[75] "Relocated Tustin Whole Foods debuts," *Orange County Register*, August 29, 2007

[76] "Vendor Certification Programs: Improving Food Safety Through Performance Tracking," *Food Safety Magazine*, July 2006

In: U.S. Food Marketing: Issues and Changes ISBN: 978-1-61761-731-7
Editors: Walter J. Kinkoff

Chapter 2

BREAKING DOWN MARKET BARRIERS FOR SMALL AND MID-SIZED ORGANIC GROWERS

Alida Cantor and Ron Strochlic

ACKNOWLEDGMENTS

We are very grateful to the many individuals and organizations that have worked to make this research possible, particularly the USDA Agricultural Marketing Service for providing funding through Cooperative Agreement 12-25-A-4823 and Adam Diamond of the USDA Agricultural Marketing Service for his assistance with all aspects of the research.

Many thanks to Paige Culver, CIRS intern, who provided invaluable support with her comprehensive analysis and write- up of the survey results. We are very grateful to Lisa Kresge, CIRS research associate, for her support in survey design and editing, and to Gail Wadsworth and Kate Snow for their editorial assistance. We would also like to thank Margaret MacSems for assistance conducting interviews with Spanish-speaking growers.

Many individuals and organizations supplied indispensable assistance with survey design and dissemination, including Claudia Reed and Jane Baker, CCOF; Tracy Lerman, Organic Farming Research Foundation; Marcy Coburn, Ecological Farming Association; Erin Barnett and Guillermo Payet, Local Harvest; Ryan Galt, UC Davis; and Aubrey White, UC Davis. Thanks as well to the Agriculture and Land-Based Training Association (ALBA), the Eco-Farm Conference, CCOF and the Community Alliance with Family Farmers for their help in organizing group discussions.

Thanks to the many key informants who shared their insights and knowledge, including: Andrew Angus, Switch Farm Marketing; Benny Fouche, University of California Cooperative Extension; Brett Melone, Agriculture and Land-Based Training Association; Carolyn Dimitri, USDA Economic Research Service; Cathy Greene, USDA Economic Research Service; Claudia Reed, CCOF; David Runsten, Community Alliance with Family Farmers; Ellie Rilla, University of California Cooperative Extension; Fred Thomas, CCOF Going Organic; Gail Feenstra, University of California Agricultural Sustainability Institute; Holly Givens, Organic Trade Association; Janet Caprile, University of California Cooperative Extension; Karen Klonsky, University of California Davis; Katherine DiMatteo, Organic Trade Association; Lydia Oberholtzer, USDA Economic Research Service; Mario Moratorio, University of California Cooperative Extension; Mark Gaskell, University of California Cooperative Extension; Mark Lipson, Organic Farming Research Foundation; Richard Molinar, University of California Cooperative Extension; Shermain Hardesty, University of California Small Farm Center; Steve Quirt, University of California Cooperative Extension; and Thomas Nelson, Capay Valley Grown.

Finally, special thanks to the farmers and buyers who took the time to enlighten us about their experiences with organic farming and marketing. Without their input this study could not have been completed.

EXECUTIVE SUMMARY

With growing consumer demand for organic food in the U.S., the organic sector has enjoyed robust rates of growth in recent years. Nonetheless, growth in organic production and farmland is not keeping pace with demand. Organic farming was initially considered a promising marketing niche for small and medium growers. While that has been true for some growers, many have also experienced considerable challenges with respect to marketing. While competition from larger growers is clearly an important factor, previous research conducted by the California Institute for Rural Studies has identified a range of additional marketing challenges facing small and mid-sized organic growers. This research sought to obtain in-depth information about these challenges through interviews and surveys with growers, buyers and experts familiar with the organic sector in general and organic marketing in particular. The research findings identify a broad range of challenges and – more importantly – a range of recommendations for addressing these challenges. We hope the findings outlined in this study contribute to making organic agriculture a more viable marketing channel for small and mid-sized growers.

The research focused on small and mid-sized organic farmers in California, primarily those producing fruits and vegetables. Areas of focus include most prevalent marketing barriers, how these barriers vary based on different farm and farmer characteristics, barriers associated with specific marketing channels, and recommendations for addressing these challenges. Methods included a review of the literature on organic marketing; an online survey of 103 farmers; three group interviews with farmers; and in-depth semi-structured interviews with 35 farmers, 22 buyers and 21 key informants.

Extent & Types of Marketing Challenges

Over 80% of growers in the study reported marketing challenges, with approximately 40% citing marketing as a major challenge. Volume (too much or too little) was the largest specific marketing challenge, as reported by 84% of survey respondents. Additional challenges include obtaining organic price premiums (66%), locating and accessing markets (65%), competition (55%), lack of pricing information (47%) and difficulties meeting buyer requirements (37%).

Challenges Based on Farm Characteristics

Marketing challenges varied based on farm size, with larger farms more likely to report competition as a challenge and smaller farms citing issues of staff time associated with marketing. Crop type affected marketing challenges somewhat, with survey respondents growing orchard crops, berries and vegetables reporting the most challenges and those growing diversified crops reporting the fewest problems. Survey respondents selling wholesale reported the most marketing problems, while farmers selling at farmers' markets reported the fewest problems. Length of time farming was not a significant factor in the survey. All Latino farmers interviewed cited major marketing challenges, particularly with respect to language barriers.

Challenges of Scale

Interviewees identified issues associated with scale as the most significant challenges facing small and mid-sized organic and conventional farmers, specifically competition with larger growers and foreign imports, consolidation and regulation. Issues specifically affecting organic farms included the growth of the organic sector, certification issues and competition with other claims including "local" and "sustainable." Successful small and mid-sized organic farms are utilizing a strategy of values-based marketing to address these challenges. They are emphasizing the values that make their farms unique and are competing on these values, rather than low prices, where they cannot compete. These successful farmers are employing creative marketing systems involving strong personal relationships, as well as the use of Internet communications.

Marketing to Wholesale, Retail and Institutional Buyers

In-depth interviews with distributors, marketers, directors of marketing cooperatives and institutional buyers shed light on the challenges of working with small and mid-sized organic farmers from a buyer's perspective. These challenges include price, transaction costs, quality, appearance, food safety, consistency, volume and availability, knowledge of business, and communication. Buyers, as well as farmers and key informants, were very interested in seeing the development of new intermediaries and distributors to address these issues, such as cooperative marketing systems, distribution hubs to aggregate product from multiple farmers and sympathetic distributors that specialize in working with smaller-scale farms.

Marketing Directly to Consumers

Direct marketing methods, including farmers' markets, Community Supported Agriculture (CSA), farm stands and pick-your-own, were described as important marketing strategies for small and mid-sized organic farms. While potentially more lucrative, interviewees also described a number of challenges to direct marketing, including geographic location, language barriers and ability to sell at profitable farmers' markets. CSAs were described as one of the most promising and profitable avenues of direct marketing. Some farmers are experimenting with new models of "bundled CSAs" involving multiple farms. Successful farmers are utilizing a diversity of direct and wholesale marketing channels that complement one another as a means of reducing risk.

Key Recommendations

Interviews with farmers generated a series of recommendations for addressing these challenges. Farmer interest in specific recommendations was then gauged via an online survey. The following are the recommendations eliciting greatest interest from survey respondents:

Addressing food system scale and improving values-based marketing

1. Improving food safety regulations to account for farm size and marketing channels
2. Providing more money and promotion for campaigns encouraging local food
3. Simplifying or reducing organic certification process for very small-scale farms

Improving marketing to wholesale, retail and institutional buyers

1. Developing a comprehensive database of farmers and buyers to improve local sourcing
2. Supporting the development of alternative distribution channels, including consolidation hubs and marketing cooperatives
3. Developing a reusable box system or reducing packaging requirements

Improving direct marketing

1. Providing grants or loans to improve direct marketing and value-added production
2. Increasing enforcement of labeling such as organic or "no-spray" at farmers' markets
3. Simplifying direct marketing regulations, such as those for farm stands

INTRODUCTION

The organic sector has experienced dramatic growth in the U.S. over recent years. Organic food sales have grown at the extremely rapid rate of 15 to 20 percent per year, skyrocketing from $3.6 billion in 1997 to $21.1 billion in 2008 (Dimitri & Oberholtzer, 2009). Previously found only at natural foods stores or farmers' markets, organics can now easily be found in mainstream supermarkets and big-box stores and are purchased by a wide range of consumers.

Despite this rapid growth in organic sales, the growth in organic production and organic farmland is not keeping pace. Organic farms in the U.S. are not producing enough to keep up with the robust demand for organic products, sometimes causing supply shortages that make for an uncertain market. Over half of manufacturers surveyed by the Organic Trade Association reported that a lack of supply has prevented them from selling more organic products (OTA, 2006). Increasingly suppliers are turning to imported organics to make up for the difference.

Given the clear demand for organic foods, why are more farmers not entering organic production? Previous research shows that marketing is one of the primary challenge s identified by farmers exiting organic production. Lack of access to markets and the inability to obtain organic price premiums were cited overwhelmingly as barriers to organic farmers' success (Strochlic & Sierra, 2007). The growth in consumer demand is not necessarily translating into increased marketing opportunities for small and mid-sized organic farmers.

Many studies have shown that marketing is a major challenge faced by small and mid-scale organic farmers and have recognized the need for more research on ways to address organic marketing challenges (Agboh-Noameshie

& Al-Ajjawi, 2004; Dimitri & Richman, 2000a; Kang, Broome, Chaney, & Swezey, 2005; Strochlic & Sierra, 2007; Walz, 2004). Growing consumer demand for organic produce is clearly documented as well (Dimitri & Lohr, 2007; Dimitri & Oberholtzer, 2009; Oberholtzer, Dimitri, & Greene, 2005). However, fewer studies have taken a detailed look into organic marketing from both growers' and buyers' perspectives.

This study examines marketing barriers and challenges from the perspectives of both growers and buyers. The goals are to better understand marketing challenges affecting small and mid-sized organic farms in California and to identify program and policy recommendations to help these growers more successfully market their products. By learning more about the challenges and potential solutions, future policies and programs can be strengthened in order to increase the feasibility of organic agriculture as a viable option for small and medium growers.

Focus of Study

The study focuses on small and mid-sized farms[1], since marketing for these farms is markedly different than for large-scale farms. Smaller farms tend to rely on a more diverse range of marketing channels, particularly direct marketing, which has a unique set of challenges (Kambara & Shelley, 2002; Oberholtzer, Dimitri, & Greene, 2005). Within the category of small and mid-sized organic farms, there are sub-categories of farmers, such as new farmers and minority and immigrant farmers that face specific challenges. This study examines specific challenges of different marketing methods as well as challenges facing different types of farms and farmers.

The geographic scope of this study is limited to California for several reasons. California has the most organic farms in the country, at 2,026 farms in 2007 and represents 41% of total U.S. organic acreage (Oberholtzer, Dimitri, & Greene, 2005). It also leads the nation in direct marketing, used predominantly by smaller-scale organic farms (Kambara & Shelley, 2002).

This study focuses mainly on farms producing fresh fruits and vegetables for a number of reasons. The majority of organic sales are fresh produce, which represented 39% of the total $13.8 billion of organic sales in 2005 (OTA, 2006). At the same time, marketing systems for other organic products such as dairy, commodity crops, etc. are very different and were ultimately deemed outside the scope of this study.

Research Questions

This study sought answers to the following questions. What are the most prevalent marketing barriers faced by small and mid-size organic farmers in California? How do these issues vary by factors such as farm size, crops, region and proximity to markets? What are the particular marketing issues facing minority, new and immigrant farmers? Why do only half of organic growers participate in direct marketing, despite its higher earning potential? What can be done to encourage and facilitate greater participation in direct marketing channels among small and medium organic growers?

METHODS

This study utilized a diverse array of methods, including a literature review, key informant interviews, grower interviews, buyer interviews, grower group interviews and a grower survey.

Literature Review

A literature review was conducted to identify key marketing issues facing small and mid-sized organic growers. The literature review addressed the context of organics, organic markets, supply chains, direct and local marketing and organic wholesale distribution.

Key Informant Interviews

Interviews were conducted with 21 key informants. Key informants were selected for their expertise in issues pertaining to smaller-scale farms, organic farms and/or organic marketing. They represent a diverse range of perspectives including UC Cooperative Extension farm advisors; researchers from government, university and nonprofit sectors; and representatives of organizations working with small and/or organic farms. Interviews lasted from 30 minutes to one hour and were conducted by telephone, with the exception of three interviews conducted in person. Key informants were asked a series of open-ended questions about the organic sector, marketing challenges, root causes of marketing challenges, and recommendations to address these challenges.

Buyer Interviews

Interviews were conducted with 22 buyers along the organic supply chain, including distributors, institutional and retail buyers and directors of marketing cooperatives. Buyers were identified via a snowball sampling method and were chosen to provide a representation of the full spectrum of the supply chain. Buyer interviews, which were conducted by telephone, lasted 20 minutes to one hour. The interviews focused on buyers' experiences working with small and mid-sized organic farms, perceptions regarding key barriers in buying from small and mid-sized organic farmers, and recommendations to make it easier to work with smaller-scale organic farms.

Farmer Interviews

A total of 35 organic farmers were interviewed. Farmers were identified using snowball sampling and were chosen to represent a diverse range in terms of size, length of time farming, location, marketing channels and immigrant and minority status. Farmer interviews focused primarily on fruit and vegetable producers. Five of the 35 interviews were conducted in Spanish. Farmer interviews lasted from 20 minutes to one hour and were conducted by telephone. The interviews focused on the extent to which marketing is a challenge, most pressing issues, challenges associated with different marketing channels, and recommendations for addressing these issues.

Group Interviews with Growers

Because facilitated discussions can elicit different information than one-on-one interviews, three group interviews were held with farmers. All were held in conjunction with other farmer meetings. Group interviews included two focus groups, one in English and the other in Spanish. A large group discussion with approximately 60 participants was also conducted with organic farmers attending a regional meeting of the Community Alliance with

Family Farmers (CAFF) and CCOF in the Central Valley. The group interviews elicited grower perceptions about pressing marketing challenges and recommendations for programs or policies to improve their marketing opportunities.

Farmer Survey

An online survey of farmers was conducted to test receptivity and interest in recommendations developed via the interviews and focus groups. The survey was disseminated via email through organizations working with organic farmers, including the Ecological Farming Association, Local Harvest, CCOF and the Organic Farming Research Foundation. The survey generated 122 responses, of which 103 were valid cases used for analysis (the remaining 19 were not farmers or did not complete the survey). The survey was in English only. Because of the dissemination method of emails sent through partner organizations, response rate is unknown. The survey took approximately 10-15 minutes to complete and included a combination of multiple choice, Likert-scale, and open-ended questions about principal marketing challenges and interest in recommendations for addressing these challenges.

Literature Review: Background on Organic Marketing

The Organic Marketplace

Organic agriculture in the United States has enjoyed soaring popularity as organic foods have moved into mainstream markets. What began as a fringe environmental movement during the first half of the twentieth century gained momentum as a health-related movement in the 1960s and 70s and has more recently become popular amongst a broader audience (Dimitri & Richman, 2000a; Guthman, 2004). The organic industry has seen immense sales growth since the 1990s, increasing by about 20 percent every year and topping $21 billion in sales in 2008 (Dimitri & Greene, 2002b; OTA, 2006).

Organics are seen as a huge opportunity by the food marketing and retail sector, with firms of all sizes hastening to jump on board (Harris, 2006).

Historically organics were predominantly found in natural foods stores and coops, but today consumers are demanding more variety and higher quality produce from their conventional grocery stores and supermarkets, which are increasingly offering organic products (Dimitri & Oberholtzer, 2009; Greene et al., 2009; Oberholtzer, Dimitri, & Greene, 2005). The consumer base has also expanded, with two-thirds of all Americans buying organic at least occasionally. Despite common perceptions of organic consumers as predominantly white and affluent, demand crosses lines of income and ethnicity (Dimitri & Oberholtzer, 2006; Stevens-Garmon, Huang, & Lin, 2007; Tropp, Ragland, & Barham, 2008).

Organic Supply Chains

Once harvested from the farm, organic food can follow several supply chains, including wholesale, direct to retailers or institutions, or directly to consumers. Wholesale products can be fresh or processed and can go through different combinations of brokers, distributors, processors and/or manufacturers before reaching a store shelf or dining establishment (Dimitri & Richman, 2000a). Organic food also can be sold directly to stores, restaurants or institutions such as hospitals or schools, or directly to consumers via farmers' markets, Community Supported Agriculture (CSA), farm stands and pick-your-own operations. Regardless of which path the food takes, organic integrity must be assured at each step and all intermediaries must comply with organic standards (Dimitri & Greene, 2002a).

Organic farmers tend to rely on a broader range of marketing outlets than conventional farmers (Oberholtzer, Dimitri, & Greene, 2005). In the 2004 National Organic Farmers' Survey, 80% of organic vegetable growers reported selling directly to consumers. Direct marketing methods tend to be more common among organic and smaller farms as well as among younger, beginning and foreign-born farmers (Kambara & Shelley, 2002; Walz, 2004).

The past 20 years have witnessed a shift in outlets where organic food is purchased. Most organic food was purchased at natural foods stores through the 1990s. However, by 2005, independent natural foods stores represented under 25% of organic sales, while mass- market channels, including supermarkets, mass merchandisers and club stores became the largest retail outlets for organic food products (OTA, 2006), which are now available at nearly three out of four conventional grocery stores (Greene, 2006).

At the same time, direct sales of locally produced organic food are also growing. Despite a shift towards larger marketing outlets, the smallest organic farms have maintained a foothold via direct sales, with direct- market demand for local organic produce in almost every region of the country (Grow & Greene, 2007; Kremen, Greene, & Hanson, 2004) and farmers' markets and CSA programs rapidly gaining in popularity. The USDA reports over 5,000 farmers' markets across the nation in 2009, almost triple the early 1990s (USDA Agricultural Marketing Service, 2009). Similarly, the first CSA program in the U.S. began in 1985. CSAs today are estimated at over 2,500 farms (Payet, 2009).

Organic Farmland

Despite robust growth in sales, land in organic production still represents only about one- half of one percent of total U.S. farmland. Domestic farmland is not meeting the demand for organics and suppliers are increasingly importing organic products from foreign markets to make up for the difference (Greene et al., 2009). While California leads the nation in organic production with about 40% of sales, only 1.4 percent of farmland is certified organic (ERS, 2009; Klonsky & Richter, 2007; Strochlic & Sierra, 2007). Furthermore, although organic acreage is increasing in California, the number of farms has remained constant or even decreased over time, indicating concentration of the organic sector and a loss of smaller-scale farms (Klonsky & Richter, 2007). Previous research has identified marketing challenges as an important factor contributing to exit patterns among small and mid-scale organic farmers (ERS, 2009; Klonsky & Richter, 2007; Strochlic & Sierra, 2007).

Challenges for Small and Mid-Sized Organic Producers

The landscape of organics has changed markedly over time. The USDA implemented the National Organic Standards in 2002 in response to growing confusion and concerns regarding the integrity of organic claims. The codification and certification of organic practices meant that direct relationships between producers and consumers were no longer necessary to ensure the validity of organic claims (Dimitri & Richman, 2000a). That organic label made it easier for larger manufacturers and retailers to sell organic products, which in turn spurred the entry of larger organic farms and resulted in increasing concentration of the organic sector (Edberg, 2004). As such, a number of observers believe that the National Organic Standards had negative impacts on market among smaller organic farms (Grow & Greene, 2007). As Dupuis (2006) explains, "Even though the market for organic food may be expanding, smaller organic farmers are increasingly unable to gain access to the mainstream buyers that represent an increasingly large portion of the growing market."

Current Challenges in the Organic Marketplace

One issue affecting the continued growth of the organic sector as a whole is a lack of consistent supply. Over half of organic manufacturers have reported that short supplies of certified organic products have restricted their company from selling more organics (Dimitri & Oberholtzer, 2008; Gogoi, 2006; Knudson, 2007; OTA, 2006). While scarcity assures higher prices for organics, it also makes retailers and manufacturers uneasy about going into such an insecure business (Dimitri & Richman, 2000a). Businesses make up for supply shortages by importing organic products. In 2004, 38% of organic distributors and processors imported some or all of their organic products, indicating that there is significant room for domestic organic production to expand (Greene et al., 2009).

Another issue facing organics is growing interest in food attributes such as "sustainable" and "local." Consumer demand for locally grown food is growing, often with a preference for local *and* organic when possible (Hartman, 2009; Pirog, 2003). The impacts of increased demand for local food are unclear (Greene et al., 2009). On the one hand, local food that is not certified organic is typically cheaper than certified organic and thereby presents competition. On the other hand, many smaller organic farms market their product locally, thereby benefiting from growing demand for local food. Unverified labels such as "natural" and "no-spray" may also compete with organic.

Research Findings

Farm and Farmer Characteristics

Farm and farmer characteristics are presented for the 35 farmers interviewed and the 103 survey respondents included in the analysis.

Years farming

Interviewees. The farmers interviewed reported farming an average of 12 years with a median of seven years.[2] One-third (34%) had been farming for five years or less, one-third (31%) between 6 and 15 years and the remaining one-third (34%) for over 15 years.

Survey. Survey respondents had farmed a mean of 17 years and a median of 13 years, ranging from less than one year to 70 years. Twenty eight percent (28%) had farmed five years or less, 31% had farmed for 6 to 15 years, while the remaining 41% had been farming over 15 years.

Acreage

Interviewees. Farmer interviewees reported farming an average of 48 acres, with a median of 11 acres. Almost half (49%) farmed 10 acres or less, while 26% had between 11 and 50 acres and 26% were farming over 50 acres.[3] Acreage ranged from less than one acre to 300 acres.

Survey. Survey respondents reported a mean of 308 acres and a median of 16 acres, with a range spanning from less than an acre to 10,000 acres.[4] Almost half (45%) farmed less than ten acres, 29% farmed between 11 and 50 acres and 24% reported over 50 acres.

Crops

Interviewees. The majority of interviewees (80%) primarily produce diversified vegetables and fruits. Several also raise animals. Of the remainder, four farmers (11%) were orchard farmers, two (6%) were primarily growing flowers and one was primarily involved in animal production.

Survey. Vegetables were grown by 65% of respondents, followed by orchard crops at 57%.[5] Berries were grown by 35%, livestock and animal products by 34% and nursery products by 23%. Sixteen percent (16%) had pastureland and 16% grew "other" products.[6] Twelve percent (12%) reported diversified operations and another 12% grew grain and/or hay.

Organic certification

Interviewees. Of the 35 farmers interviewed, 32 (91%) were certified organic, while three (9%) used organic practices but were uncertified.

Survey. The survey was open to all farmers who self- identified as organic. Interestingly, over half (52%) were farming using organic practices, but were not certified organic.

Marketing channels

Interviewees. Farmer interviewees reported a diverse range of marketing channels. The most common, used by 77% of farmers, was direct sales to retailers and/or restaurants. The next most common was farmers' markets, as reported by 69% of farmers, followed by wholesale at 66%. CSA strategies were used by 45% of farmers. Twenty nine percent (29%) had farm stands while 20% reported agri-tourism or pick-your-own operations.

Most growers reported multiple marketing channels. The majority (86%) cited two to four marketing channels (29% used two channels, 31% used three and 26% used four). Three farmers (9%) were using five or more channels. Only two (6%) reported just one marketing channel.

Survey. Survey respondents reported a diverse range of marketing channels as well. The most common channel, used by 70% of farmers, was direct sales to retailers and/or restaurants, followed by farmers' markets (62%). Forty five percent (45%) sold through farm stands or pick-your-own, 44% sold wholesale and 20% sold directly to institutions.

Most survey respondents used multiple marketing channels. The majority (81%) reported two to four marketing channels (31% used two channels, 30% used three and 20% used four). Only 7% used five or more marketing channels, while 12% used just one marketing channel.

Overview of Marketing Challenges

Extent of marketing challenges

Of the survey respondents, 39% felt marketing was a major problem, while 46% said it was "a little" or "somewhat" of a problem and 15% reported it was not a problem. Similarly, approximately half (46%) of interviewees reported that marketing is a major problem, while 34% felt it was somewhat of a problem and 20% felt it was not a problem (see Figure 1).

Grower interviews indicate that marketing is essential to farm success. As a farmer explained, "It's number one because if you can't sell it, it doesn't matter—everything else depends on it." Similarly, a Cooperative Extension farm advisor responded, "Growing is not *the* challenge. Anyone can grow crops. The marketing is what makes or breaks it." A second farm advisor pointed out, "Although there are other problems, they could be solved if markets were better and if people got more money for their produce."

Interviews also showed that marketing problems span a wide range, from farmers who found marketing "by far the biggest problem" to those who "are able to sell everything, but it's a lot of work." Farmers who did not report marketing challenges attributed their success to factors including years of experience and high local demand. As a farmer explained, "There are relatively few farmers here…we're community heroes in a way." Another farmer commented, "Until recently we had more demand than supply. Marketing is easy when you sell out."

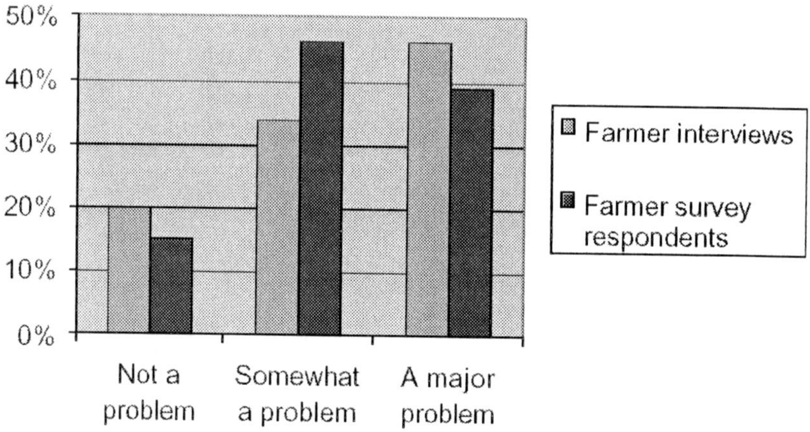

Figure 1. Extent to which marketing is a problem

Specific marketing challenges

Survey respondents were asked to indicate the extent to which specific marketing challenges, including getting organic price premiums, volume (too much or too little), accessing markets, competition, meeting buyer requirements, and pricing information affected their farms (see Figure 2). The principal challenge cited was volume (too much or too little), as reported by 84% of respondents. Those selling at farm stands and/or you-pick operations reported the most trouble with volume, at 91%, while those selling wholesale had the fewest volume cha llenges, at 78%.

Getting organic price premiums was the second most challenging problem, as reported by 66% of survey respondents. Those selling wholesale cited the most problems getting organic price premiums, at 79%. Organic price premiums had the smallest effect on those selling direct to institutions, at 65%.

Locating and accessing markets was a problem for 65% of respondents. Farms selling direct to institutions reported the most trouble locating and accessing markets, with 70% citing that as a problem. Conversely, farm stands and/or you-pick operations the fewest problems, at 54%.

Over half (55%) of respondents cited challenges associated with competition. Farms with CSAs reported the most competition, at 73%, with farm stands and/or you-pick farms reporting the fewest issues (44%). Competition was the only specific issue varying significantly by farm scale: whereas only 35% of those farming ten acres or less reported competition as a problem, 63% farming 11 to 50 acres and 74% farming over 50 acres reported issues with competition.

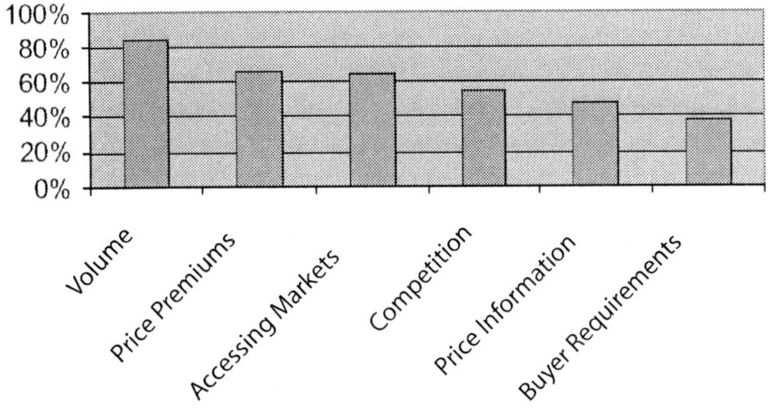

Figure 2. Specific marketing challenges

About half of the respondents (47%) cited lack of price information as a challenge. Lack of price information was most common among farms selling direct to institutions, with 58% reporting problems. Those least affected were selling at farmers' markets, at 42%.

The least challenging of the specific marketing issues was meeting buyer requirements, with only 37% reporting that as a challenge. Those selling direct to institutions reported the most challenges, with 53% citing it as an issue, while those selling at farmers' markets had the least trouble at 35%. New and smaller-scale farmers reported slightly higher rates of challenges meeting buyer requirements.

Marketing organic products

Almost half of survey respondents (42%) reported that they had sold organic product as conventional due to a lack of marketing opportunities for organics. A number explained that they could not sell their product as organic because they were not certified, despite following organic production practices. This problem was not however limited to uncertified farmers, as similar percentages of certified and uncertified organic farmers reported selling organic product as conventional. Farmers selling wholesale (54%) were more likely to report this. Larger farms were also more affected, with 46% of farms over 50 acres and 48% of farms of 11 to 50 acres selling organic as conventional, compared with 33% of farms with 10 acres or less.

One-quarter (25%) of survey respondents also reported that marketing challenges prevented them from expanding their organic acreage. This varied by organic certification status. Of respondents not certified organic, only 15% said that marketing had prevented them from expanding organic acreage, the case for 28% of respondents farming solely certified organic acreage. Farms with some certified organic acreage and some uncertified acreage were by far the most affected, with 45% reporting that marketing challenges had prevented them from expanding organic acreage.

Marketing challenges based on farm characteristics

The study examined differences in marketing challenges based on factors including farm size, crops, length of time farming and minority/immigrant status.

Farm size

Competition was the only specific challenge associated with farm scale. Whereas only 35% of those farming ten acres or less reported competition as a problem, 63% farming 11 to 50 acres and 74% farming over 50 acres cited concerns with competition. Grower interviews also indicated challenges associated with the amount of time required for marketing, particularly among very small-scale operations. As a farmer noted, "It's a big challenge for a small farm like us. It's just the two of us and if we're out at the farmers market we're not here doing farming."

Crops

In the survey, marketing challenges varied somewhat by crop. Those growing orchard crops reported the highest rates of marketing challenges (89%), followed closely by berries (85%) and vegetables (84%). Respondents producing diversified crops reported the fewest marketing challenges, at only 64%.

Marketing channels

Survey respondents using wholesale channels reported the highest rates of marketing challenges, with 93% reporting marketing as a problem. Farms selling direct to institutions, retailers and restaurants tied with the second highest rates of challenges, at 90% each. Those selling at farmers' markets had the lowest rates of marketing challenges (85%). Later sections of this report provide an in-depth look at specific issues affecting wholesale and direct marketing.

New farmers

Survey data did not indicate significant differences in challenges based on years farming. However, a number of interviewees noted that marketing is more difficult for newer farmers. One farmer called the first year "especially difficult" since it takes time to develop connections. Another beginning farmer explained, "We're not inherently good at marketing, we didn't do it before we started this farm. We'd grown things before, but marketing was a new skill." Conversely, a more experienced farmer explained that the key to success was that "thirty years of experience has made [marketing] not a challenge anymore."

Minority and immigrant farmers

Growers in the survey were not asked to report their ethnicity. The interviews includ ed six Latino farmers (17% of interviewees), with five interviews conducted in Spanish. [7] A focus group was conducted in Spanish.

The interviews and focus group found that ethnicity is associated with marketing challenges. All (100%) of the Latino interviewees reported marketing problems, compared with 75% of non-Latino farmers. While this sample is not large enough to make definite conclusions, it is notable that all Latino interviewees reported significant marketing barriers. Language barriers were reported as the biggest concern, since that prevents Latino growers from accessing markets and participating in direct marketing.

Food System Scale and Values-Based Marketing

One of the main issues facing small farmers, whether organic or conventional, is that the U.S. food system's policies and infrastructure are better-suited for larger farms, which makes it difficult to succeed as a small farmer. Organic agriculture originally offered a niche market that sheltered smaller farms from the pressures of the conve ntional market. However, with the rapid growth of the organic sector, small and mid-sized organic farmers now face many of the same pressures as conventional farms, in addition to specific challenges related to organic production and marketing (Guthman, 2004).

Many smaller farmers are tackling these challenges by utilizing a values-based marketing approach. They are finding ways to differentiate themselves by building personal relationships with their customers, telling their unique stories and communicating the values embodied in their food. Organic production practices are one part of this values-based marketing, along with other values such as small-scale, local and family farming.

Challenges: Scale and food system structure

The study respondents emphasized that many of the marketing problems facing small and mid-sized organic farmers are more a function of scale than organic production and marketing. As a key informant explained, "These problems aren't unique to organic. They are unique to smaller farms." A farmer added, "It's not an issue so much of organic versus non-organic, it's small producers versus large producers." Competition, consolidation and regulation were identified as key issues negatively affecting all small farms, organic or not.

Competition with large farms

Farmers and researchers cited competition with larger farms and foreign imports as one of the principal barriers affecting small farms. A researcher explained, "Small and medium farms are competing against imports and large producers. Costs per unit are higher for small producers, so it's hard to compete." Another researcher pointed out that competition depresses organic prices, since "large farms produce monocrops or are highly specialized, which allows them to lower their costs of production. This makes the prices of the small producers appear higher." Larger growers push down the price premiums for organic farmers of all sizes (Oberholtzer, Dimitri, & Greene, 2005). Other research has also identified competition as a major challenge. Respondents to the Fourth National Organic Farmers' Survey cited competition with large farms as their main marketing challenge, followed by competition with imported organic products (Walz, 2004).

Import competition

Organic farms of all sizes face competition from foreign imports, which comprised about 12% of the U.S. organic market in 2002 and are growing rapidly (Grow & Greene, 2007; USDA Foreign Agricultural Service, 2005). While imports have fueled the growth of the organic sector as a whole, they present a concern for domestic organic farmers. Since organic farming is labor intensive, countries with lower labor costs can produce more cheaply, driving domestic prices down (Grow & Greene, 2007). Fortunately for smaller farmers, research indicates that they experience less competition with imports because of direct marketing and consumer demand for locally grown organic products (Greene et al., 2009; Oberholtzer, Dimitri, & Greene, 2005). That was confirmed by the farmer survey, in which competition was reported as a much smaller problem for farmers under 10 acres than farms over 50 acres. Nonetheless, import competition is a significant concern particularly affecting mid-sized growers.

Farm advisors working with organic growers were particularly concerned with import competition. As one observed, "The prices are too low because they're competing with China and South America, which have lower production costs. We have higher labor and environmental standards. This means we have very safe food, but it's expensive." Another added that "Farmers here are competing with those in Mexico, where the cost of living and input costs are cheaper. It's more expensive to produce food locally than to ship produce from other countries. Globalization of organic produce is threatening smaller farms." An organic nut farmer cited personal experience with import competition, noting that "We sold directly to Trader Joe's for a while, but then this year they got a shipment from Moldova. We can't compete with that."

Consolidation in the food system

Consolidation of buyers and farms results in a loss of infrastructure for smaller-scale farms, as well as loss of power in decision- making (Hendrickson & Heffernan, 2002; Howard, 2009; Lyson, Stevenson, & Welsh, 2008). As a researcher exp lained, "Consolidation and concentration have taken away the trucks and other intermediaries that serve small and medium sized farms. The market system doesn't serve them anymore." Small farmers find it difficult to compete in a system with fewer, more dominant players. A farmer explained: "As the organic business gets more industrial, it looks more and more like conventional agriculture. Larger buyers want to work with larger producers. Larger buyers are gobbling up smaller buyers, who would work with smaller farms and larger farms are gobbling up smaller farms." A researcher pointed out that this is especially problematic for organics, since "Organic distribution infrastructure is actually *more* concentrated than conventional agriculture."

Regulatory compliance

Several interviewees observed that smaller-scale farms have much less political influence compared with larger farms, resulting in regulations favoring larger farms. "We need to make over the rules and state regulations," suggested a farm advisor. "They were created to benefit larger growers…they're really onerous and bureaucratic. It makes it hard for small growers to get through the process." A farmer spoke of difficulties complying with regulations, noting that "Larger companies can afford it, but smaller ones can't."

Small farms face difficulties complying with regulations that do not take their creative and diversified systems into account. Farmers frequently cited concern that new legislative activity regarding food safety and leafy greens will hamper their production and marketing systems by forcing all farmers, small and large, to comply with the same set of strict regulations. As one farmer noted, "The industry is making everyone wear the same shoe, even though it's by everyone wearing the same shoe that we got into trouble in the first place." Farmers producing meat also cited trouble complying with regulations requiring that animals be slaughtered in certified USDA facilities, since there are very few facilities that work with smaller farms.

Access to information

Many growers cited challenges associated with limited access to information about small-scale organic production and marketing. Critics have argued that land- grant university research and extension activities have served the needs of larger conventional farmers and that smaller and organic farms have largely been absent from the university research agenda (Hightower, 1973). Very few farmers reported using Cooperative Extension and university resources, although many said they would like to if they offered information relevant to small-scale organic farms.

Lack of organic price information was another information gap. Pricing information is typically tracked through wholesale markets. However, since organics are often sold directly to consumers or retailers, it is difficult to track organic prices (Sok & Glaser, 2001). The organic sector would clearly benefit from more university research and outreach on production and marketing systems, as well as more thorough and accurate pricing information—both basic governmental supports provided for larger and conventional farms.

Specific challenges to small and mid-sized organic farms

While many marketing challenges affect all small farms, some are unique to organic. These include the growth of the organic sector, certification issues and the relationship between organic and competing attributes.

Growth of the organic sector

No longer just a small niche market of farmers' markets and natural food stores, the organic sector has grown exponentially to include many large-scale farms and buyers. For many small and mid-sized organic growers, the rapid growth of the organic sector has resulted in unstable price premiums, loss of niche markets and increased competition (Hanson, Dismukes, Chambers, Greene, & Kremen, 2004). A long-time organic farmer described the changes over 30 years of farming: "It's a different world now. The quality demands are much higher, the marketplace is different, the capital needs are more. In the early days, there was not a lo t of organic product, so there was high demand for ours." Another farmer added, "It used to be that the organic marketing system was friendly to small farmers, because that's what the system was – organics were small. But now, big businesses have taken over the landscape. It's going the way of industrial now."

Organic certification issues

In 2002 the USDA implemented the National Organic Standards to standardize production and other practices and increase consumer confidence in organic claims (Strochlic, 2005). These standards have generated heated debate within the organic sector. Many farmers view the paperwork and certification costs as prohibitive. As one explained, "Organic certification involves lots of hoops and paperwork. You need a good price premium to justify the increased costs." A farm advisor cited certifications costs as a significant barrier for immigrant growers, explaining that "The cost of certification has been a big challenge for our growers, especially the Southeast Asian growers. We had a dozen certified organic by the state program, but when the national regulations went into effect, we lost all of them."

Competition from "local" and "sustainable"

Over recent years, "local" and "sustainable" foods have gained considerable traction. Popular books and articles have informed consumers that while organic certification guarantees a set of production practices, it does not ensure that food was produced on small-scale, local, or family-owned farms (Pollan, 2006; Roberts, 2009). However, it is rarely pointed out that local has no explicit scale component, nor does it address production practices.

While the issue is often framed as "local versus organic," many interviewees emphasized that the two can and should work together. As a farmer noted, "organic is important, but it's not everything—the local and organic combination is the best." Similarly, a researcher explained that, "The press has put this emphasis on local versus organic and turned it into a debate, without recognizing that it can be both. But the press has misinformed people. It's not a dichotomy, lots of local is organic."

A number of buyers expressed strong interest in local food. As a dining services director explained, "We focus more on local than organic. I think the movement is going places other than organic. Certification isn't important to me if I can see what their practices are and tell growers' stories." Another food services provider added, "Our mandate is local, not organic. [Organic] distributors buy from huge farms. It's not from small family farms who are trustees of the land. So we go for local."

However, a distributor saw this as problematic, pointing out that, "There's an over-romanticization of local right now, a lack of clarity around what local means. It's about whole systems, not just local." As a researcher added, "Big corporations can be local. But that doesn't necessarily mean they're good."

Opportunities: Values-based marketing

Due to competition from larger farms, foreign imports and competing attributes, organic certification alone is no longer an adequate marketing strategy for many small farms. The organic label conveys information about growing practices only. Farmers must therefore convey other values that consumers increasingly seek, including scale, local, sustainable or family farmed (Conner, 2004). As a key informant explained, "If you are local and diversified and a family farm, organic can be part of a market strategy, but it is not a strategy in and of itself." One farmer explained that differentiation is necessary, because "there's no difference in labeling between those doing 'barely-organic' for the price premium and those doing a good job. Those with integrity get undercut by others doing it cheaper."

The farmers interviewed noted that the ability to convey values other than organic is the most important aspect of marketing, requiring a personal and relationship-based style. As one farmer explained, "If you are faceless, nameless and don't have a personal presence, you won't survive." Farmers described their marketing as "Very personal, based on conversations and connections." In this style of marketing, telling the story is as important as the ability to produce food. As one farmer described: "It's easy to grow a lot of food, but people want to know about you, your life story, the history of your farm.... You have to make a product that is intimate and relationship based."

Competing on values

Interviewees stressed the importance of competing on attributes other than low price. As a farmer recommended, small farmers should "Explain why someone should buy your stuff for your price. If they don't buy it, someone else will. Never compete on price as a small farmer. They either want organic, high quality produce and are willing to pay for it, or the y're not."

Another farmer added, "I see a lot of farmers trying to compete on price. Don't bother. Price is a losing proposition. Compete on quality and values such as environment, social justice." The same farmer pointed out that "You have to tell your story if you're competing on values, or it doesn't work."

Many farmers also emphasized the importance of competing on quality. A farmer described how she has "built a brand by never selling anything that wasn't 'A' quality," which has given her another important value besides price on which to compete.

Community-based advertising

Given the importance of values and personal relationships, many farmers rely on a strong community presence to establish credibility and communicate their values, rather than traditional advertising. Community-based advertising methods include write-ups in local newspapers and magazines, participating in community events and benefit dinners, hosting school tours, attending conferences and inviting people to the farm. As a farmer explained, "Creating goodwill and good relations in the community is important. Good PR is good marketing." Another added, "I'm always marketing – I'm always talking about my farm to friends and community members."

Creative and adaptive marketing

Along with maintaining a personal presence and building strong relationships with customers, farmers stressed the importance of being creative, flexible and diversified. Since there is no one-size- fits-all model for smaller farms, marketing must be tailored to each farm. As a farmer noted, "I recommend that small farms don't do what big farms are doing – find a little niche and do it well. Be flexible and ready to change quickly if big farms start doing it." A farmer with a thriving agri-tourism operation described what works for her: "We're too small for any regular accounts. So we stay diversified and do agri- tourism. We have to be creative because we're small."

Online communications

Online communication and social networking are increasingly valuable marketing tools for many farmers. Websites, email newsletters and social networking sites are popular ways for growers and customers to connect. Farmers felt that a user-friendly website was important, and for some, "imperative." Several described getting email addresses at farmers' markets and developing lists to advertise CSA shares and farm events. A farmer explained that an email newsletter is a useful way "to connect with people and tell them what you're up to." A number of farmers mentioned success using Local Harvest, a website with a searchable directory of small farms. These virtual technologies are helping farmers establish a personal presence and build closer connections with consumers.

Recommendations: Scale and values-based marketing

The interviewees offered a number of recommendations to address scale-related challenges and improve values-based marketing for small and mid-sized organic farmers. Recommendations are listed in order of farmer interest.

Improving food safety regulations

Interviewees recommended changing food safety regulations to better suit small farms and assistance to help small growers comply with regulations. Modifying food safety regulations was the most popular recommendation in the survey, with fully 91% of respondents reporting that regulations accounting for farm size would be helpful (66% "very helpful," 25% "somewhat helpful"). An additional 84% would find funding for smaller farms to comply with food safety regulations and develop food safety protocols helpful (56% "very helpful," 28% "somewhat helpful").

Encouraging local purchasing

Respondents emphasized the importance of encouraging consumers to purchase from small and local farmers, recommending that more money and promotion put into "Buy Local" campaigns. An impressive 95% of survey respondents answered that this would be helpful (59% "very helpful," 36% "somewhat helpful"). Respondents also recommended that public agencies directly support local organic farms through purchasing, with 81% of survey respondents answering that policies supporting local and organic purchasing for public agencies would be helpful (53% "very helpful," 28% "somewhat helpful").

Organic certification options

A large percentage (83%) of survey respondents felt that a simplified certification process for the smallest-scale farmers would be helpful (61% "very helpful," 22% "somewhat helpful"). Additionally, 77% of respondents felt that different labeling options for "barely organic" and "beyond organic" farms would be helpful (39% "very helpful," 38% "some what helpful").

Consumer education on organics

Farmers also stressed the need for increased consumer education around what organic certification does and does not mean, with 86% reporting that this would be helpful (48% "very helpful," 38% "somewhat helpful"). Given concerns about food safety, 90% of farmers called for greater consumer education regarding organics and food safety; particularly to dispel consumer perceptions that organic food is unsafe (50% "very helpful," 40% "somewhat helpful").

Assistance in storytelling and digital communications

As websites, computers and online social networking continue to serve as important marketing tools for smaller-scale farmers, some farmers, particularly immigrant or limited-English farmers, require more technical support in digital communications. In terms of general communications assistance, 79% of survey respondents were interested in help with developing marketing materials and "telling their story" (35% "very interested," 44% "somewhat interested"). An additional 76% express interest in assistance with digital communications, Internet marketing and developing websites (29% "very interested," 47% "somewhat interested").

General support and research

Interviewees expressed interest in more general support for smaller organic farms, including organic-specific production and marketing research and Cooperative Extension outreach, which 79% would find helpful (32% "very helpful," 47% "somewhat helpful"). An additional 78% expressed interest in low-cost consultants or advisors to help with marketing and business planning (40% "very helpful," 38% "somewhat helpful"). Interviewees also called for more support for farmers transitioning to organic, such as the existing "Going Organic" program.

Improving price information

Several farmers and buyers cited difficulties making marketing decisions due to a lack of organic price information. That is borne out by the survey, where 78% of respondents reported that access to improved organic price information would helpful (37% "very helpful," 41% "somewhat helpful"). The findings indicate that organic price information would likely be more useful to farms selling wholesale.

Support for immigrant and minority farmers

Latino farmers reported particularly high rates of marketing challenges. Language barriers are key challenge, given the important role communication plays in values-based marketing. Latino growers also cited challenges associated with limited computer skills. These challenges indicate a need for programs assisting farmers to ensure they are reaching minority and limited-English growers. Organizations could help connect Latino growers to Latino consumers (who have shown high interest in organics) or could provide more assistance accessing markets and using computers. Creative solutions should also be considered, such as pairing Americorps volunteers with limited-English farmers to assist with direct marketing.

Marketing to Wholesale, Retail and Institutional Buyers

Wholesale marketing channels include distributors, processors and manufacturers. In this study, retail and institutional buyers are included as well.[8] Retail and institutional buyers represent a marketing channel somewhere between direct marketing and wholesale. They are similar to wholesalers in that sales involve larger quantities than direct marketing and there is an intermediary between the farmer and the end consumer. For farmers, the main benefit of selling wholesale (to distributors or directly to retailers or institutions) is moving large volumes of product at once. The main drawback is that prices are lower than direct sales. Farmers must take factors such as time, price and volume into account in determining whether it makes sense to sell wholesale.

Challenges: Wholesale marketing

Interviews with buyers identified a number of barriers associated with working with smaller farms, including price, transaction costs, logistics, quality, appearance, food safety, consistency, volume, availability, and knowledge of business, markets, and communications. This confirms and builds on other research examining wholesale and direct-to-retailer marketing barriers for smaller farms (Starr et al., 2003; Zajfen, 2008). The severity of these challenges may explain why growing demand for organics is not translating into increased marketing opportunities for small and mid-sized farms. As the market for organics expands in mainstream retail formats, it does not expand for small-scale farms because it is difficult for them to access these larger channels. Challenges are described in detail below, roughly ordered by importance to buyers as determined by the interviews.

Price

Price is one of the primary barriers for buyers sourcing from small and mid-sized organic farms. Organics command a price premium and smaller farms must cover higher per-unit production costs. From the point of view of many buyers, the prices considered fair and necessary by small-scale organic farmers are too high. "Price is the number one challenge in selling organics," commented one distributor. "It needs to be within a certain range and organics [from small farms] are out of the price range." A wholesale distributor who works with smaller farms pointed out that customers who buy wholesale produce "have access to bigger suppliers with lower prices, so we have to keep prices low enough to compete." Several buyers noted that while they wanted to support the higher production costs of smaller farms, the price point was higher than their business could handle.

Price is particularly an issue in institutiona l settings, such as school, university, or hospital cafeterias. Whereas local organic food simply commands a higher shelf price in a retail setting, cafeterias often have a fixed menu at a certain price. "It's hard for us to buy organic because of the pric e point. Our meals have a set price and it's all- you-can-eat, so it is hard to pay extra money for organic options," explained one cafeteria director. Another cafeteria buyer added, "We currently only buy from local organic farms if the price is close to conventional prices. Otherwise we can't buy it." A distributor explained, "Most customers have stated that they'd like to buy organic, but price is a concern and organic is mostly out of reach. Customers, especially institutions, have a hard- line price point and strict budgets." Even buyers who want to support small organic farms have a hard time doing so within their required price points.

Transaction costs and logistics

High transaction costs add to the issue of high prices. Per unit transaction costs are lower when buying large quantities from one supplier and costs increase purchasing smaller amounts from multiple suppliers. Buyers trying to keep transaction costs low therefore have an incentive to buy large quantities from a small number of farms, one of the major barriers preventing buyers, particularly larger ones, from working with smaller farmers. "It takes the same amount of time to coordinate two boxes as it does 200 boxes or a truckload," explained one distributor. Another estimated that it costs a minimum of $100 in time, fuel and logistics every time a truck stops at a farm, no matter the size. As a small distributor noted, "Costs as a small distributor are major. It's a higher cost per unit than if we had a huge amount like a larger distributor."

Another added, "Transportation is more difficult when you can't fill a whole pallet. The costs of fuel and trucking go much higher." Transaction costs are also an issue when farmers sell directly to retailers or institutions. As a cafeteria director expla ined, "It takes more time to order food and locate it when we source it locally. It's a big automated system and local produce is like a wrench in that system."

Quality, appearance and packing

Buyers expressed concerns regarding the issues of quality, appearance and packing. Wholesale produce must be clean with no cosmetic defects. It must be packed properly in the right boxes, with proper post-harvest handling procedures and cannot be overripe.

Wholesale distributors require attractive, clean produce to sell to their customers, who are accustomed to particular standards. Distributors noted that it is "difficult to get a clean, consistent pack from small growers." An institutional cafeteria buyer added, "Quality and appearance are issues with small and orga nic. We've rejected a lot more produce [from smaller farms] on the basis of quality."

Another distributor noted issues with post-harvest handling: "With smaller growers we have trouble with post- harvest handling and cooling. Even selling locally, the product needs to have some shelf life for a few days." Another distributor said it is important to ask small farmers, "Do you have the cooling facilities, the storage facilities? Post-harvest handling is so important. Being a good farmer is more than having a green thumb."

Packaging and boxes are additional challenges in working with smaller farms, since each type and amount of produce requires a different box. Farmers unfamiliar with selling wholesale often do not pack produce in the proper boxes. A distributor explained, "A lot of farmers try to save money by not buying the right boxes. But then the boxes collapse and are rejected so it costs more money. You need the proper packaging."

Buyers also stressed that it can be difficult working with smaller farmers accustomed to direct marketing quality standards. "Labeling, counting, aesthetics – there are different standards. What works for the farmers' market doesn't work for customers buying wholesale," one distributor explained. Another added, "Farmers don't understand. For a farmers' market it goes right to the consumer and they eat it right away. We need product that can sit in cold storage a few days."

Another distributor felt that smaller growers have quality problems because they are unwilling to cull product: "It costs more to produce everything for them. That's fine, but in some aspects they value everything too much. They won't cull enough and the quality goes down." Buyers emphasized that farmers must think about where the product is going and pack it accordingly.

Food safety

Food safety is a huge issue for farmers, buyers, food service providers and consumers. Some buyers require farmers to conduct audits or carry liability insurance, which can be a significant barrier for small farmers. "Food safety is a huge issue, so we need some kind of audit to minimize risk," one distributor explained. Smaller farmers can have a difficult time meeting complex standards. Nonetheless, one buyer felt that organic farmers had an easier time meeting food safety standards because they already meet legal requirements for organic certification.

Consistency and availability

Wholesale distributors require consistent availability to ensure adequate supply. Smaller farmers described difficulties working with wholesalers who need a commitment to a certain volume at a certain time. From the distributors' point of view, it is difficult working with small farms that cannot guarantee availability. As one distributor described, "It's more volatile. Availability is less consistent and less predictable." For institutional buyers, dealing with seasonal availability challenges is a big learning curve. As a cafeteria director explained, "If you order cherry tomatoes, they might run out, there might be a frost or broken equipment. With a big produce company they just find it for you from somewhere, whenever."

Knowledge of business and markets

A number of buyers emphasized that a better understanding of wholesale markets would help growers plan and make better decisions. As a wholesaler commented, "Farmers should know the market conditions, pricing, competition – we don't want someone just growing something without knowing whether it's a good thing to grow." Another added, "Even small farms need to understand the markets. They should be planting crops based on what they need to make."

On the opposite end, buyers sourcing from smaller organic farms want their end customers to have a better understanding of the farmers' context. A distributor who sells to restaurants and stores stressed that "Chefs and retailers need to understand that small growers can't provide large uniform quantities."

Another wholesale distributor felt he could play an important role in terms of educating his customers and "explaining so people can understand. The customers want parsnips, but the ground is frozen. Weather and other factors require understanding." Developing relationships and common understandings between buyers, farmers and end customers is an important starting point for overcoming other obstacles.

Communication

Buyers stressed that one of the most important things they need from farmers is frequent and clear communication. Buyers need to know what a farmer has, in what quantity and when it is available. They must know these things early, rather than when the product becomes ripe. "If I'm selling one of your products, I need to hear from you a few times a week. You need to call me back. Don't leave me hanging," explained one distributor. Another emphasized, "Number one is killer communication! It's difficult to over-communicate with us." Communicating about the product itself is important as well, especially for local and organic produce. Explained one buyer, "We always want more info from growers—who they are, what they're up to. If we're asking customers to pay more for food, we need to show them why and what the story is." It can be "difficult to get information to the end consumer," added an institutional buyer, so having as much information as possible from the farmer about the product is critical.

Farmers recognized the importance of good communication. "The more interaction with your [wholesale] customers the better. We talk daily with some of our customers, just keeping a dialogue," explained one farmer who sells wholesale. Another farmer suggested "going to talk with people about what you grow, bringing samples. Let them see and taste it." As in direct marketing, small and mid-scale organic farmers selling wholesale are asking buyers to pay more for values. It is therefore important to convey the values and story behind the food.

Opportunities: Values-based distribution chains

Consumer demand for organic food from small local farms is quickly growing and institutions, retailers and manufacturers are working hard to find ways to meet this demand. Farmers, intermediaries and end buyers committed to supporting small-scale local organic farms are working together to form "values-based supply chains" (Dimitri & Richman, 2000b; DuPuis, 2006). One of the strongest messages in this study was the need for alterna tive distribution channels and intermediaries who work specifically with smaller farms. Interviewees from all sectors supported the development of such alternative distribution channels. Potential options include 'sympathetic distributors' that specialize in working with small- scale and organic farmers; cooperative marketing systems; and distribution hubs that serve as consolidation points.

Farmer interest

Several farmers expressed strong interest in creating cooperative distribution systems and distribution hubs to aggregate product. One described a group of local farmers who are currently "looking for warehouse space to act as a distribution hub." Farmers are organizing to form agricultural marketing cooperatives such as the Capay Valley Farm Shop and the Central Coast Agricultural Cooperative. However, farmers noted that despite interest and enthusiasm, "It's hard to figure out how it will make money because it takes time to organize." Another farmer added, "We don't know where to begin or how to get help with it. I only have a limited amount of time to dedicate to it and the other farmers are in the same boat." One farmer suggested that "more micro grants, micro lending programs and small grants for feasibility studies" would help farmers start cooperatives.

Buyer interest

Since values-based supply chains rely on commitment from all players, buyers play an important role. Buyers in the study emphasized that they strongly value local and organic small- scale farms and wanted to buy food that was in line with their values. An organic produce distributor commented, "I don't really like large monoculture organic farms. Our mandate of working with small farms continues." Another distributor explained that, "We are really founded in values. Our mission is to encourage sustainable agriculture, work with local producers and get produce to businesses." A food service provider envisioned her dining facility as "a living laboratory, where students come learn how to live sustainably and responsibly." Although buyers identified significant challenges working with small farms, they also believed it was important to do so, which is encouraging for promoting values-based supply chains.

Buyers were interested in distribution systems that would ease the burdens of working with many smaller farms. "It would be helpful to have consolidation points where a distributor could pick up from multiple farmers," suggested one buyer. A distributor described how helpful it was to work with a small farm that consolidates produce from other local farms, so the distributor's truck only has to stop once. An institutional buyer remarked, "It behooves farmers to get together and organize." By working in groups, farmers can strengthen their ability to sell directly to larger buyers such as institutions or grocery stores. A buyer for a large grocery chain suggested that farmers "form co-ops to make a bigger supply. If a co-op buys from many farmers, we can buy from them." Efficiency was a motivator for the creation of aggregation systems. One buyer pointed out that if each farmer delivers produce separately, it results in "lots of empty trucks driving back. That's not sustainable! Distribution hubs would be more efficient." One challenge is that cooperatives and distribution hubs would entail another layer of intermediaries, which would likely increase costs.

Need for more research and support

Key informants saw a great deal of possibility in alternative distribution systems, but stressed the need for "cost-effective ways to aggregate product among many producers." Researchers felt that distribution networks are "ripe for analysis to figure out what could make them more effective." Experts emphasized the need for more research, financial and technical support to develop alternative distribution systems and cooperatives.

There are a number of examples of alternative distribution systems for smaller-scale and organic farms. Some are retailer- initiated programs driven by consumer demand. For example, La Montanita Co-op, a consumer foods cooperative in New Mexico, started a distribution network to source products from regional small and mid-sized organic producers. Other models are entrepreneurial ventures such as Red Tomato, a Massachusetts-based business that coordinates a regional network of farmers, truckers, wholesalers and customers to source from small and mid-sized local farms. Other programs are organized by nonprofits, such as the Growers Collaborative and ALBA Organics, which operated by the California-based Community Alliance with Family Farmers (CAFF) and Agriculture and Land Based Training Association (ALBA) respectively. Nonetheless, on a national scale there are relatively few models, despite high interest on the part of farmers, buyers and advocates. Because there is so much interest, it is important to avoid reinventing the wheel or repeating past mistakes by studying existing models and providing adequate support and information for new ventures.

Recommendations: Improving wholesale marketing opportunities

Buyers, farmers and key informants had a number of recommendations to improve wholesale marketing opportunities for small and mid-sized organic farms. Recommendations are presented based on farmer interest.

Database of farmers and buyers

There is currently no centralized mechanism for connecting local farmers with buyers. Buyers wishing to source from local farmers have no way of finding them or knowing what they have for sale, while farmers do not know which buyers are interested in local organic products. In order to facilitate connectio ns, respondents recommended an online searchable database listing farmers and buyers. Databases exist for a few regions,[9] but a more centralized statewide or regional database would be a valuable resource. Fully 84% of survey respondents expressed interest in a database of this nature (41% "very interested," 43% "somewhat interested"). Online ordering systems with real-time information regarding availability of produce were seen as helpful by 61% of respondents (24% "very helpful," 37% "somewhat helpful").

Development of alternative distribution systems

Farmers, buyers and key informants expressed high support for the development of alternative distribution systems including sympathetic distributors, distribution hubs or aggregators and marketing cooperatives. Fully 80% of respondents would find the development of cooperatives or aggregate distribution systems helpful (39% "very helpful," 41% "somewhat helpful"). There is a significant need for research and technical support for such systems. Funding in the form of grants or loans will be needed to provide startup capital for new distributors or cooperatives.

Boxes and packaging

Several farmers and buyers cited boxes and packaging as significant marketing challenges for small and mid-sized growers. Boxes can cost a dollar or more each and different boxes are often required for each type of product. This quickly turns into a major expense and creates a lot of waste. One recommendation was for small farmers to order boxes cooperatively to achieve economies of scale. Farmer advocacy organizations could help facilitate this approach, which is already happening in some communities.

Farmers and buyers also expressed interest in reusable box systems. In the survey, 75% of respondents answered that a reusable box system would be helpful (36% "very helpful," 39% "somewhat helpful"), while another 75% felt that reducing packaging requirements would be helpful (44% "very helpful," 31% "somewhat helpful").

Exchanges between buyers and farmers

Several interviewees expressed interest in "cultural exchanges" between farmers and buyers to promote understanding and improved relationships. Farmers would visit stores, restaurants and cafeteria kitchens, while buyers would visit farms. As a cafeteria director explained, "It would be helpful to have visits both ways. We could see what each other does and understand each other's barriers. Farmers might give us new ideas." A farmer who had hosted a field tour noted, "Having buyers out to the farm was important. They tasted things and saw the farm and liked it and then we got better prices." In the survey, 65% of respondents felt that gatherings and field days between buyers and farmers would be helpful (21% "very helpful," 44% "somewhat helpful").

Buyers supporting farmers

Buyers and handlers wishing to assure a steadier supply of organic product can help by taking a more proactive role, helping farmers with organic transitions, post-harvest handling, micro- loans and crop planning (Dimitri & Oberholtzer, 2008; Trim, 2007). For example, Whole Foods Market has a Local Producer Loan Program to help capitalize small farms. A buyer described how his distribution company assures a supply of high-quality produce by providing farmers with technical assistance in pest management and post-harvest handling. A marketing cooperative held organic transition workshops so the cooperative would have enough organic supply. A wholesale distributor used crop planning to ensure that farmers would have a market and the distributor would have supply: "We work with farms to figure out a schedule of who grows what and where, to make sure we have a steady supply." In the survey, 54% of farmers answered that loans, support and technical assistance from wholesale distributors would be helpful (22% "very helpful," 32% "somewhat helpful").

Farmer education on selling wholesale

Given buyer challenges working with smaller farmers, it would be beneficial to educate farmers on how to sell wholesale. There are several existing manuals that address these topics (Dufour, 2008; Feenstra, Ohmart, & Chaney, 2003), which could be expanded to workshop formats. In the farmer survey, 71% of farmers felt that workshops on selling wholesale would be helpful (16% "very helpful," 55% "somewhat helpful"). Several buyers expressed interest in teaching farmers about quality and post-harvest handling standards. As a distributor noted, "We're working on a grant to teach farmers about wholesale pack expectations." Yet another added, "I would like to teach a course on how farmers can work with the distribution and retail marketplace."

Marketing Directly to Consumers

Direct marketing, including farmers' markets, CSA, roadside stands and pick- your-own is an important marketing channel for smaller-scale organic farms, allowing farmers to earn more by avoiding intermediaries. Direct marketing has soared in popularity with growing interest in local food and there are currently over 5,000 farmers' markets and over 2,500 CSA programs in the U.S. (USDA Agricultural Marketing Service, 2009; Greene et al., 2009; Hartman, 2009; Payet, 2009). As a researcher observed, "The proliferation of direct marketing opportunities—farmers' markets, CSAs—has gone very well and is allowing for a lot of robust marketing activity at a small scale."

Challenges: Direct marketing

Farmers and key informants identified a number of challenges associated with direct marketing, including geographic location, language barriers and access to profitable farmers' markets.

Geography

Interviewees cited geographic location as both an asset and a challenge. Previous research indicates that demand for organic produce varies regionally and is strongest near urban areas and universities (Kremen, Greene, & Hanson, 2004). As a farm advisor in the San Francisco Bay Area explained, "Marketing isn't a problem here. Everyone sells out. We are near San Francisco, with a huge affluent foodie-oriented population."

More often, location was referred to as a challenge, particularly for those in more rural areas or near communities with low demand for organic. As one farmer explained, "The Bay Area is 300 miles away, but we can't sell organics up here because it's too rural and conservative. So we load up trucks and go to the Bay." This farmer did not even advertise her produce as organic in her hometown until recently: "We only started telling people it was organic in the last few years—we wanted to be part of the community. It's considered wacky and expensive and people have negative perceptions around it."

For farmers lacking an adequate local market, transportation to more profitable markets can be expensive and difficult. At a focus group, farmers described transportation time and fuel costs as a major challenge. One very rural farmer also noted that, "Accessing markets is a big challenge because of our remote location. We're too remote for a CSA."

Nonetheless, rural does not always equate to low demand for organic. A farmer in a small community with high demand for local organic products claimed that, "I'm lucky to be where I am. There is a huge market."

Language barriers

Latino farmers cited language barriers as their biggest challenge, particularly with respect to direct marketing. Smaller farmers must communicate directly with customers to engage in values-based marketing. For those with limited English, it is impossible to convey these values to an English-speaking customer base. As a farmer described, "Language is the biggest issue. It makes it impossible for some farmers to do farmers' markets or other direct marketing. They can't understand Spanish and we can't understand English."

Farmers with limited English also have a hard time eliciting feedback from customers. As a farmer explained, "If I spoke better English I could talk with customers and know what to bring to farmers' markets." Another farmer noted that language is a problem because "Customers want to know more about how we manage our crops and there's no communication between the consumer and the farmer."

Respect is an additional issue affecting Latino growers working at farmers' markets, many of whom have reported a perceived lack of respect on the part of market managers.

Access to profitable farmers' markets

Farmers' markets are the most common format of direct marketing. They facilitate direct interaction between consumers and farmers, create lively public spaces, improve access to fresh food for consumers and create markets for local farmers (Gillespie, Hilchey, Hinrichs, & Feenstra, 2007). They are generally considered more profitable because revenues go directly to the farmer. However, research shows that farmers' markets are not necessarily profitable for farmers once the value of their time and labor costs are included (Hardesty & Leff, 2008), given the costs associated with harvesting for market, driving there, spending the day, packing up unsold produce and driving back. As one farmer commented, "It would be nice to do more direct marketing, but we don't do farmers' markets because of the time commitment." Another echoed, "Farmers' markets are too much time and energy."

Farmers were quick to note that some farmers' markets, typically in wealthy areas, can be very profitable and well worth the time. Profitable markets have some combination of higher traffic, customers willing to support higher prices, and larger sales volumes. However, as one farmer pointed out, "Farmers' markets that move volume have very long waiting lists. The other markets aren't worth the time." Another added, "The good ones are saturated and the ones with space are not as good."

According to a key informant, "the market must be very good for it to be worth it and only a fraction of farmers' markets are that good." While recent growth in the number of farmers' markets throughout the country is typically seen as positive, it may actually contribute to the problem of unprofitable farmers' markets. A farmer described how "an over-saturation of markets in my area" has resulted in "not enough business at each." Adding more markets can dilute sales at each individual market to the point where they become financially unfeasible for farmers.

Opportunities: Direct Marketing

Community Supported Agriculture

Community Supported Agriculture (CSA), in which customers become shareholders of a farm in exchange for weekly shares of produce, began in the U.S. in the 1980s and has rapidly grown into a popular form of direct marketing. Farmers were very positive about CSAs as a marketing channel. One long-time organic farmer called CSA "the most exciting thing I see going on these days—it's a bright spot." Since CSA involves predictable sales, farmers can harvest knowing that the produce is already sold. As a farmer commented, "The farm stand is too unpredictable, it's too much waste. The CSA is easier to plan for." CSAs can also play an important role as part of a larger farm system. One farmer said that her CSA "gets good feedback, we can grow fun things and it allows for year-round employment and health insurance for our workers."

A CSA can also create a market for produce that would otherwise go unsold. A farmer noted that their CSA meant that, "I can sell produce that is too ripe for wholesale." This confirms previous research describing CSAs as an important risk management strategy for organic farmers (Hanson, Dismukes, Chambers, Greene, & Kremen, 2004).

A study of growers' experiences with CSA found that the primary disadvantage was inefficiency, including time spent with members and the diversity of crops required (Perez, 2004). However, the same study found that farmers were positive about their experiences with CSAs, which had created economic security. Hardesty and Leff (2008) found that when time and labor were taken into account, CSA was a more profitable marketing channel than farmers' markets or wholesale.

A new trend is the development of "bundled CSAs" incorporating produce from multiple farms. Several farmers with large CSAs described buying from other farms to supplement their CSAs. As one explained, "If small farms want to focus on growing and don't want to market themselves, we would be happy to sell their produce via our CSA." Another farmer explained that including produce from other farms in his CSA is advantageous because "they [the other farmers] get a better price than wholesale and we get a good product that we can't grow ourselves." This readiness to collaborate confirms previous research that even when CSA farms are seemingly in competition with one another, there is a strong sense of collaboration as well (Strochlic & Shelley, 2004).

One key informant was dubious about this approach, noting that, "People might prefer that direct relationship with the farm, so bundling might not work." However, farmers currently running bundled CSAs report positive feedback from customers, who enjoy a greater variety of products in their weekly share.

Diversity in marketing systems

Farmers stressed the importance of diversified marketing systems, using multiple channels to find the approach best suited to their situation and sell a greater percentage of their product. For example, one farm coordinated with a café kitchen to use blemished produce: "We avoid waste with the kitchen. A farm in harvest has a lot of waste, but we have an alternative – we send it to the kitchen."

Other farmers used diverse marketing systems as an insurance strategy. "We have so many things, our products are diverse enough, that we can tap into different markets and if the price falls on one, there are still others," explained one farmer. Another added, "Our key has been not putting all the eggs in one basket. I recommend not just selling wholesale or just the farmers' market. Having a mix of strategies is critical. It allows us to utilize all our stuff." Previous research has shown that even though diversified operations require more time and energy, diversified marketing strategies reduce risk, increase earnings and increase success for smaller-scale farmers (Hardesty & Leff, 2008; Park & Lohr, 2006; Tourte et al., 2006).

Recommendations: Improving direct marketing

Respondents generated a number of recommendations for programs and policies to improve direct marketing. These are presented based on farmer interest.

Grants for direct marketing and value-added production

Interviewees recommended grants or loans for farmers wishing to scale up their marketing systems or expand into value-added production, which can be capital intensive. While government value-added production grant programs already exist, interviewees recommended expanding these opportunities. This generated positive interest in the survey, with 79% of respondents expressing interest in grants for direct marketing and value-added production (47% "very helpful," 32% "somewhat helpful").

Enforcement of labeling at farmers' markets

Certified organic farmers recommended increased enforcement of rules regarding labeling and product claims at farmers' markets. Because certified organic farmers invest time and money in maintaining their certification, it is problematic when "Those who are not organic say 'spray-free' without having any certification. They have cheaper prices. Organic is more expensive than 'spray- free' to produce, so people don't buy [organic]." Another farmer claimed that, "People say it's organic when it's not. It's a fraud. We do a lot of work to be organic certified." Seventy five percent (75%) of survey respondents felt that greater enforcement of product claims at farmers' markets would be helpful (50% "very helpful," 25% "somewhat helpful").

Simplifying regulations

Several farmers and key informants recommended simplifying regulations regarding direct marketing, particularly at farmers' markets and farm stands. Several local governments, including Contra Costa and Sacramento Counties, have recently passed legislation facilitating direct marketing. Of survey respondents, 76% would find simplifying regulations helpful (43% "very helpful," 33% "somewhat helpful").

Farmers raising livestock noted that regulations and infrastructure hamper their success as well. Animals must be slaughtered in USDA-approved facilities; however, there are few that work with small producers. Farmers described driving four or more hours to reach a slaughter facility. Simplifying regulations and/or developing more local and mobile meat processing facilities would assist smaller producers and those wishing to diversify into livestock production.

Facilitating bundled CSAs

Farmers expressed interest in bundled CSAs as well. Although such CSAs are typically coordinated by one farm, there are cases in which they are coordinated by a nonprofit or cooperative entity. Seattle's Pike's Place Market ran a bundled CSA program with multiple farmers; the farmer cooperative Capay Valley Vision is creating a bundled CSA of regional farmers; and Placer County piloted a collaborative CSA project. When asked about support for development of bundled CSAs, 71% of survey respondents answered that it would be helpful (27% "very helpful," 44% "somewhat helpful").

Workshops and toolkits for direct marketing

Several farmers expressed interest in approaches such as CSA and agri-tourism, but did not have enough information to implement a program on their own farm. Technical assistance, workshops, toolkits and how-to manuals providing information on specific marketing methods would be helpful for farmers interested in learning more about direct marketing. In the survey, 73% of farmers felt that toolkits or workshops on direct marketing would be helpful (21% "very helpful," 52% "somewhat helpful").

Encouraging EBT and WIC at farmers' markets

Several farmers expressed interest in more use of EBT (food stamps) and WIC coupons at farmers' markets. Along with increasing food security, this could encourage new farmers' market shoppers and increase business at farmers' markets in less wealthy communities. It could also be a good strategy to connect Latino farmers with Spanish-speaking customers. The strongest support for this strategy was expressed at the Spanish- language focus group. In the survey, 56% of respondents felt this would be helpful (20% "very helpful," 36% "somewhat helpful").

Analysis and Recommendations

Detailed descriptions of recommendations for addressing scale and improving values-based marketing, improving wholesale marketing opportunities, and improving direct marketing are provided throughout this study. This section analyzes these recommendations, looking first at recommendations specifically relevant to organic farming, examining farmer interest in solutions, and dividing recommendations by relevance to different audiences.

Organic-specific Issues and Recommendations

Many of the marketing barriers affecting small and mid-sized organic growers are a function of scale and subsequently affect all smaller growers. Nonetheless, organic farmers face a number additional barriers not faced by other smaller farms, including the following:

- The need to recoup higher production costs.
- Unstable demand and prices for organic products.
- Competition with competing attributes such as "local," "sustainable" and "no-spray."
- Consumer concerns regarding food safety of organic products.
- Consumer confusion regarding the fact that orga nic does not necessarily mean small, local or family- farmed.
- Limited access to organic price information.
- Limited university research and outreach for organic production and marketing.
- Low demand for organic products in certain regions.

Marketing barriers facing small and mid-sized organic growers may be best exemplified by the fact that 41% of survey respondents reported selling organic products as conventional due to a lack of marketing opportunities for their organic products.

An issue of particular importance is organic certification, which raised the some of the strongest organic-related commentary in the study. Respondents discussed the consequences of the national organic certification system: by creating one standardized label with no nuances or differentiation, the national standard means that small local organic farms must compete with very large organic farms and organic farms overseas. Additionally, the expenses of organic certification can prove overwhelming for the smallest of farms. While some respondents criticized organic certification, others felt strongly that using organic certification is essential to preserving the integrity of organics.

To address certification issues, respondents recommended finding more options to distinguish between "barely organic" and "beyond organic," and creating a simplified and cheaper organic certification process for the smallest of farms. Finding a way to help transitioning organic farmers get price premiums during the three-year transition period, or subsidizing the transition to organic, would be helpful as well. Other organic-specific recommendations included increased consumer education about the meaning of organics and about organics and food safety; increasing purchasing of organics by public agencies; and improving university research and outreach for organic farmers.

Farmer Interest in Solutions

Survey respondents were asked to rank the importance of four broad categories of assistance they would find useful: (1) direct marketing; (2) wholesale distribution; (3) information, communication and technology; and (4) regulations and certification. As seen in Figure 3, direct marketing, regulations/certification and information/communication/technology were all ranked highly. Assistance with wholesale distribution was ranked as less important.

Category	**Importance (scale of 1-10)**	**% Responding "Important"[10]**
Direct Marketing	7.4	92%
Regulations and Certification	6.9	86%
Information, Communication and Technology	6.6	90%
Wholesale Distribution	4.5	65%

Figure 3. Importance of broad categories of solutions

The survey also elicited farmer perceptions of specific solutions generated through interviews with farmers, buyers and others. As seen in Figure 4, the most popular recommendations are related to direct marketing and regulations/certification. Of the top-ranked recommendations, three are related to consumer education while two are related to food safety. The least popular recommendations are related to wholesale marketing, likely due either to the fact that only 44% of respondents sell wholesale or to lack of interest in wholesale due to the challenges involved. Notably, all recommendations received interest from over 50% of farmer respondents.

Recommendation		Importance (1-10)	% Responding "Important"
1.	Food safety regulations accounting for farm size, marketing methods	7.9	91%
2.	More money and promotion for buy-local campaigns	7.6	95%
3.	Simplified organic certification process for smallest-scale farmers	7.2	83%
4.	Funding to help smaller farms comply with food safety regulations	7.0	84%
5.	More consumer education on organics and food safety	6.9	90%
6.	More consumer education on the meaning of organic	6.8	86%
7.	Policies supporting local and organic purchasing for public agencies	6.7	81%
8.	Farmer-to-farmer networking opportunities for sharing information	6.6	85%
9.	Increasing grants for direct market-ing and value-added production	6.3	79%
10.	Databases to connect growers and buyers based on products	6.2	84%
11.	More enforcement of "no-spray" advertising in farmers' markets	6.2	75%
12.	Simplifying regulations for direct marketing	6.0	76%
13.	Development of marketing cooperatives or distribution systems	5.9	80%
14.	Low-cost consultants to help with business planning and marketing	5.9	78%
15.	Reduced packaging requirements	5.9	75%
16.	Improved information about organic prices	5.8	78%
17.	Different labeling for "barely organic" vs "beyond organic" farms	5.8	77%
18.	Help with developing marketing materials and "telling the story"	5.7	79%

(Continued)

	Recommendation	Importance (1-10)	% Responding "Important"
19.	More organic research and Cooperative Extension assistance	5.5	79%
20.	Reusable box system to reduce box costs	5.5	75%
21.	Retail-format farm stands or stores that sell local produce	5.3	74%
22.	Help with cutting-edge technology, Internet marketing and websites	5.2	76%
23.	Development of "bundled CSAs"	4.9	71%
24.	Toolkits or workshops for farmers' markets, CSAs, agri-tourism	4.7	73%
25.	Gatherings and field days between distributors and farmers	4.3	65%
26.	Online ordering systems to reduce need for middle buyers	4.2	61%
27.	Workshops on how to sell wholesale	3.8	71%
28.	Encouraging use of EBT (food stamps) and WIC at farmers' markets	3.8	56%
29.	Loans, technical assistance from wholesale buyers	3.8	54%

Figure 4. Grower interest in potential solutions

Farmer Information Sources

The survey elicited information on farmer preferences for obtaining information about marketing. The most popular sources of information were other farmers and Internet resources, each preferred by 70% of survey respondents. Buyers, distributors and customers were the next most preferred source of information at 56%, followed by email (55%) and farmer organizations (52%). Additional sources of information were magazines and newspapers (40%), workshops and classes (37%) and conferences (28%). Cooperative Extension advisors and books were cited by only 27% of respondents. Only 8% cited consultants, while 5% preferred audio books and pod casts and 3% preferred TV and DVDs for marketing information.

These findings point to the importance of farmer-to-farmer networking, which is confirmed by the fact that 85% of respondents would find increased opportunities for farmer-to-farmer networking helpful (48% "very helpful," 37% "somewhat helpful").

Recommendations by Sector

To facilitate action, recommendations are divided by relevance to different audiences, including farmers, buyers, policymakers, organizations serving growers and researchers.

Farmers

Small and mid-sized organic farmers themselves can improve their marketing opportunities by:

- Differentiating themselves from large organic and small conventional farms through direct relationships with customers and telling the story of their farms.
- Competing on values rather than price.
- Developing diversified production and marketing systems.
- Working together to access larger markets through marketing cooperatives, distribution hubs and bundled CSAs.
- When selling wholesale, focusing on quality and appearance, packing, frequent communication, and developing knowledge of the wholesale market.
- Sharing information with each other via farmer-to- farmer networks.

Buyers

Buyers can improve their ability to meet consumer demand by sourcing from small and mid-sized organic farms by:

- Educating growers on wholesale quality and business standards.
- Supporting the development of marketing cooperatives and consolidation points to aggregate product and reduce transaction costs.
- Working with growers to develop cropping plans to coordinate supply and demand.
- Supporting organic transitions through loans or technical assistance.

- Communicating with customers about the values and specific issues of buying from small and mid-sized organic farms.
- Sharing information with other buyers about the issues and opportunities involved in working with smaller-scale farms.

Policymakers

Policymakers can support small and mid-sized organic farmers by:

- Increasing support for organics via policies explicitly recognizing the social and environmental benefits of organic production.
- Expanding existing programs supporting organic agriculture, including certification cost-share, research and technical assistance and payments for conservation practices.
- Preferential purchasing of local and organic foods by public agencies.
- Expanding research and outreach to organic growers through land grant universities.
- Improving organic data collection and dissemination.
- Improving organic price information.
- Helping to connect growers and buyers through the development of online databases.
- Improving regulations regarding food safety and direct marketing to take smaller-scale organic farms into consideration.

Organizations

Organizations that work with small and mid-sized organic farms can improve marketing opportunities by:

- Educating consumers on the meaning of organics and the importance of buying locally from small organic farms.
- Supporting information transfer by facilitating exchanges between farmers and buyers and farmer-to-farmer networking opportunities.
- Helping farmers learn about and access funding opportunities.
- Consolidating existing information about farming and marketing.
- Developing a centralized database of farmers and buyers.
- Hosting workshops on marketing, food safety, working with wholesalers, online marketing, and organic transitions.
- Supporting development of distribution infrastructure, cooperatives and consolidation hubs.

- Providing more outreach and technical assistance for limited-English farmers, as well as assistance linking immigrant and minority farmers with members of those communities.

Research

Several research needs were identified in the study, including:

- Research on alternative distribution networks including cooperatives, distributors, values-based supply chains, and aggregation hubs.
- Research on organic production[11] and marketing systems specifically relevant to small and mid-sized organic farms.
- Local foods system research to identify opportunities for strengthening local demand and improving market access for small and mid-sized farms.

CONCLUSIONS

Marketing presents significant challenges for many small and mid-sized organic farms. While many challenges are related to farm scale, a number are unique to organic growers. Farmers are finding creative ways to overcome marketing challenges through values-based marketing, developing personal relationships with customers and telling the story of their farms.

Direct marketing has historically played a key role in values-based marketing for smaller-scale organic farms, and continues to do so. However, as consumer demand for locally produced organic food from smaller farms increases, wholesale, retail and institutional buyers are increasingly interested in working with smaller-scale farms. While many buyers expressed strong interest in working with small and mid-sized organic farmers, they also cited significant challenges, including price and transaction costs. These challenges create a major bottleneck in marketing for small and mid-sized organic farmers, who in turn experience difficulties in accessing these larger marketing channels. There is a strong need to develop financially viable ways to bridge this gap between smaller organic farms and larger distribution and marketing channels. The development of values-based supply chains, including marketing cooperatives, distribution hubs and sympathetic distributors could help smaller farmers access the markets that are currently accessible only to larger farms.

Farmers expressed high interest in a range of recommendations, particularly improved food safety regulations, increased consumer education, and expanded opportunities for farmer-to-farmer networking. There are many clear opportunities for farmers, buyers, policymakers, organizations and researchers to improve marketing opportunities for small and mid-sized organic farmers and increase the viability of this key sector of U.S. agriculture.

REFERENCES

[1] Agboh-Noameshie, A. R. & Al-Ajjawi, S. (2004). *Impact Assessment of Organic Farming Program in Three California Counties*. Davis, CA: UC Sustainable Agriculture Research and Education Program.

[2] Conner, D. (2004). Beyond organic: Information provision for sustainable agriculture in a changing market. *Journal of Food Distribution Research, 35*(1), 34-39.

[3] Dimitri, C. & Greene, C. (2002a). Organic food industry taps growing American market. *Agricultural Outlook,* October 2002.

[4] Dimitri, C. & Greene, C. (2002b). *Recent growth patterns in the U.S. organic foods market*. Dimitri, C. & Lohr, L. (2007). The US Consumer Perspective on Organic Foods. In M. Canavari & K. D. Olson (Eds.), *Organic Food: Consumers' Choices and Farmers' Opportunities*. New York: Springer.

[5] Dimitri, C. & Oberholtzer, L. (2006). A Brief Retrospective on the U.S. Organic Sector: 1997 and 2003. *Crop Management,* doi:10.1094/CM-2006-0921-0 7-PS.

[6] Dimitri, C. & Oberholtzer, L. (2008). *Using vertically coordinated relationships to overcome tight supply in the organic market*: USDA Economic Research Service.

[7] Dimitri, C. & Oberholtzer, L. (2009). *Marketing U.S. Organic Foods: Recent Trends from Farms to Consumers* (No. 58): USDA Economic Research Service.

[8] Dimitri, C. & Richman, N. (2000a). *Organic Food Markets in Transition.*

[9] Dimitri, C. & Richman, N. (2000b). Organic foods: Niche marketers venture into the mainstream. *Agricultural Outlook,* June/July 2000.

[10] Dufour, R. (2008). New Markets for Your Crops [Electronic Version]. *National Center for Appropriate Technology* from

http://www.attra.ncat.org/attra-pub/new_markets.html.

[11] DuPuis, E. M. (2006). Civic markets: Alternative value chain governance as civic engagement.*Online: Crop Management,* doi: 10.1094/CM-2006-0921-09-RV.

[12] Edberg, K. (2004). *The growth of the organic market: Producers' perspectives.* Paper presented at the Agricultural Outlook Forum

[13] ERS. (2009). USDA Economic Research Service: Organic Production. from http://www.ers.usda.gov/data/organic/

[14] Feenstra, G., Ohmart, J. & Chaney, D. (2003). *Selling Directly to Restaurants and Retailers.* Davis, CA: University of California Sustainable Agriculture Research and Education Program.

[15] Gillespie, G., Hilchey, D., Hinrichs, C. & Feenstra, G. (2007). Farmers' Markets as Keystones in Local and Regional Food Systems. In C. Hinrichs & T. Lyson (Eds.), *Remaking the North American food system: strategies for sustainability.* Lincoln, NE: University of Nebraska Press.

[16] Gogoi, P. (2006). Going organic: The profits and pitfalls. *Business Week,* May 25, 2006.

[17] Greene, C. (2006). U.S. Organic Agriculture. In K. Wiebe & N. Gollehon (Eds.), *Agricultural Resources and Environmental Indicators: Economic Information Bulletin No. EIB-1 6.*

[18] Greene, C., Dimitri, C., Lin, B. -H., McBride, W., Oberholtzer, L. & Smith, T. (2009). *Emerging Issues in the U.S. Organic Industry* (No. Economic Information Bulletin Number 55). Washington, D.C.

[19] Grow, S. & Greene, C. (2007). *Impacts of international organic markets on small U.S. producers.* Paper presented at the 105th EAAE Seminar, 'International Marketing and International Trade of Quality Food Products', Bologna, Italy.

[20] Guthman, J. (2004). *Agrarian Dreams: The Paradox of Organic Farming in California.* Berkeley and Los Angeles, CA: University of California Press.

[21] Hanson, J., Dismukes, R., Chambers, W., Greene, C. & Kremen, A. (2004). Risk and Risk Management in Organic Agriculture: Views of Organic Farmers. *Renewable Agriculture and Food Systems, 19,* 218-227.

[22] Hardesty, S. & Leff, P. (2008). Working Paper: *Determining marketing costs and returns in alternative marketing channels*: University of California Davis: Department of Agricultural and Resource Economics.

[23] Harris, D. (2006). Opportunities and challenges for organic from the retailing perspective. *Online: Crop Management,* doi: 10.1094/CM-

2006-0921-08-PS.

[24] Hartman. (2009). *The Healthy Eating Report*: The Hartman Group.

[25] Hendrickson, M., & Heffernan, W. (2002). Opening Spaces through Relocalization: Locating Potential Resistance in the Weaknesses of the Global Food System. *Sociologica Ruralis, 42*(4).

[26] Hightower, J. (1973). *Hard Tomatoes, Hard Times: A Report of the Agribusiness Accountability Project on the Failure of America's Land Grant College Complex*. Cambridge, MA: Schenkman Pub. Co.

[27] Howard, P. H. (2009). Consolidation in the North American Organic Food Processing Sector, 1997 to 2007. *International Journal of Sociology of Agriculture and Food, 16*(1), 13-30.

[28] Kambara, K. M. & Shelley, C. (2002). *The California Agricultural Direct Marketing Study*. Davis, CA: California Institute for Rural Studies.

[29] Kang, A. C., Broome, J. C., Chaney, D. & Swezey, S. L. (2005). *Survey of Organic Research and Extension Activities at the University of California*. Davis, CA: University of California Agriculture and Natural Resources: Sustainable Agriculture Research and Education Program.

[30] Klonsky, K. & Richter, K. (2007). *Statistical Review of California's Organic Agriculture- 2000- 2005*. Davis, CA: Agricultural Issues Center, University of California.

[31] Knudson, W. A. (2007). *The organic food market, The Strategic Marketing Institute:* Working Paper: The Strategic Marketing Institute.

[32] Kremen, A., Greene, C., & Hanson, J. (2004). *Organic produce, price premiums, and eco-labeling in U.S. farmers' markets*: Economic Research Service, USDA.

[33] Lyson, T., Stevenson, G. W. & Welsh, R. (2008). *Food and the Mid-Level Farm: Renewing an Agriculture of the Middle*. Cambrige, MA: MIT Press.

[34] Oberholtzer, L., Dimitri, C. & Greene, C. (2005). *Price premiums hold on as U.S. organic produce market expands*.

[35] OTA. (2006). *Organic Trade Association's 2006 Manufacturer Survey*: Nutrition Business Journal.

[36] Park, T. & Lohr, L. (2006). Choices of marketing outlets by organic producers: Accounting for selectivity effects. *Journal of Agricultural & Food Industrial Organization, 4*.

[37] Payet, G. (2009). Community Supported Agriculture. from http://www.localharvest.org/csa/

[38] Perez, J. (2004). *Community Supported Agriculture on the Central*

Coast: The CSA Grower Experience.

[39] Pirog, R. (2003). *Ecolabel Value Assessment: Consumer and Food Business Perceptions of Local Foods*. Ames, IA: Iowa State University.

[40] Pollan, M. (2006). *The Omnivore's Dilemma: A Natural History of Four Meals*. New York: Penguin.

[41] Roberts, P. (2009). Spoiled: Organic and Local is So 2008. *Mother Jones,* March/April 2009.

[42] Sok, E. & Glaser, L. (2001). Tracking wholesale prices for organic produce. *Agricultural Outlook,* October 2001.

[43] Starr, A., Card, A., Benepe, C., Auld, G., Lamm, D., Smith, K., et al. (2003). Sustaining local agriculture: Barriers and opportunities to direct marketing between farms and restaurants in Colorado. *Agriculture and Human Values, 20*, 301-321.

[44] Stevens-Garmon, J., Huang, C. L. & Lin, B. -H. (2007). Organic demand: A profile of consumers in the fresh produce market. *Choices: The magazine of food, farm, and resource issues, 22*(2).

[45] Strochlic, R. (2005). *Regulating organic: Impacts of the National Organic Standards on consumer awareness and organic consumption patterns*. Davis, CA: California Institute for Rural Studies.

[46] Strochlic, R. & Shelley, C. (Cartographer). (2004). *Community Supported Agriculture in California and Washington: Challenges and Opportunities*

[47] Strochlic, R. & Sierra, L. (2007). *Conventional, Mixed, and "Deregistered" Organic Farmers: Entry Barriers and Reasons for Exiting Organic Production in California*. Davis, CA: California Institute for Rural Studies.

[48] Tourte, L., Gaskell, M., Smith, R., Fouche, C., Koike, S. & Mitchell, J. (2006). *Organic certification, farm planning, management, and marketing*.

[49] Trim, K. (2007). Organic supply and demand. *Food Processing*.

[50] Tropp, D., Ragland, E. & Barham, J. (2008). *Supply Chain Basics: The dynamics of change in the U.S. food marketing environment*: USDA Agricultural Marketing Service.

[51] USDA Agricultural Marketing Service (2009). Number of Operating Farmers Markets. From http://www.ams.usda.gov/: U.S. Department of Agriculture, Agricultural Marketing Service.

[52] USDA Foreign Agricultural Service (2005). *Linking U.S. agriculture to the world: U.S. market profile for organic food products*: U.S. Department of Agriculture, Foreign Agricultural Service

[53] Walz, E. (2004). *Final Results of the Fourth National Organic Farmers' Survey: Sustaining Organic Farms in a Changing Organic Marketplace.* Santa Cruz, CA: Organic Farming Research Foundation.

[54] Zajfen, V. (2008). *Fresh Food Distribution Models for the Greater Los Angeles Region: Barriers and Opportunities to Facilitate and Scale Up the Distribution of Fresh Fruits and Vegetables*: Center for Food & Justice: Urban & Environmental Policy Institute, Occidental College.

End Notes

[1] New USDA typology defines eight types of farms, with "small farms" generating under $250,000 in yearly sales. In this study, no specific parameters of farm scale were used and respondents were instead asked to use their own definition of farm scale.

[2] This number may be low, since several farmers answered they had been farming "30 years or more" which was entered as 30 years.

[3] Percentages may not add up to 100% due to rounding.

[4] Several farmers reported pastureland acreage, which resulted in a wide range for this question.

[5] Respondents could choose more than one answer, so percentages do not equal 100%.

[6] "Other" included herbs, edible flowers and honey production.

[7] Because similar research is currently being conducted with farmers from Southeast Asia by UC Berkeley researcher Jennifer Sowerwine and others, this research did not focus on Hmong and other Southeast Asian farmers.

[8] Marketing directly to retailers, restaurants and institutions is sometimes considered a form of direct marketing. However, for the sake of this analysis, it is considered a form of wholesale marketing, since buyers act as intermediaries rather than as end consumers.

[9] Some examples include Placer County's "Placer Grown" program, Sacramento County's listing of growers, Local Harvest's online database and the Local Food Guide produced by the Community Alliance with Family Farmers.

[10] 'Important' groups the categories of 'Somewhat important' and 'Very important.' The difference between these two categories is accounted for in the weighted score from 0-10, which explains the difference between the percentage scores and the weighted scores.

[11] A number of farmers commented that in order to market successfully, they need more research support for organic production systems that will improve their ability to grow high-quality organic products.

In: U.S. Food Marketing: Issues and Changes ISBN: 978-1-61761-731-7
Editors: Walter J. Kinkoff

Chapter 3

THE U.S. FOOD MARKETING SYSTEM: RECENT DEVELOPMENTS, 1997-2006

Steve W. Martinez

ABSTRACT

Major recent developments in the U.S. food system include the increasing presence of nontraditional grocery retailers, such as supercenters and drugstores, and competitive responses by traditional grocers, such as supermarket chains. These developments have contributed to sharp increases in concentration in the grocery retail sector, changing conventional relationships among retailers, wholesalers, and manufacturers. In such a competitive domestic food market, food companies are attempting to differentiate themselves from the competition by reporting voluntary activities that demonstrate social responsibility and by more-tailored advertising campaigns and product offerings.

Keywords: Food marketing system, food manufacturing, food distributors, concentration, corporate social responsibility, competitive strategies

Acknowledgments

Appreciation is extended to Ron Larson, Kyle Stiegert, Gary McBryde, Jim MacDonald, Warren Preston, and Leland Southard for manuscript review, Elise Golan for suggestions on organization, Courtney Knauth for editorial assistance, and Anne Pearl for layout and cover design.

Summary

A main development in the U.S. food marketing system is the influx of stores that have not traditionally sold food but that increasingly offer food items. With the growing trend for eating out, traditional food retailers also face competition from the foodservice industry.

What Is the Issue?

The products and services provided by the food marketing system (which includes manufacturers, wholesalers, and retailers) account for over 80 percent of consumer expenditures on food. Developments in this system have important implications for the cost, quality, and variety of food products. Hence, agribusiness operators, policymakers, and consumers alike are interested in how competition within the industry affects industry performance. Heightened competition has contributed to consolidation and sharp increases in grocery retail concentration, changing conventional relationships within the food marketing chain. Competition has also generally put short-term downward pressure on food prices. This report examines these structural changes and selected performance dimensions of the U.S. food system over the past 10 years.

What Did the Study Find?

The study shows that nontraditional food retailers increased their share of food sales for at-home use from 17.1 percent in 1994 to 31.6 percent in 2005. These companies were able to position themselves within the food industry by creating new shopping formats that appealed to consumers and by lowering costs. The study also found that foodservice facilities (restaurants, for example) continued to increase their share of all food sales, from 46.1 percent in 1994 to 48.5 percent in 2005, by offering new products and services.

In response to competition from nontraditional food retailers and the food-service segment, conventional grocery retailers are employing their own cost-cutting and differentiation strategies. Cost-cutting tactics include supply chain initiatives such as data-sharing activities. For instance, through UCCnet, an Internet platform, food retailers and suppliers can exchange information that facilitates product delivery and reduces out-of-stock items and excess inventory. Another cost-saving strategy is to restructure operations to focus on the most profitable stores and geographic areas.

A main cost-cutting strategy of traditional grocery retailers over the past 10 years has been to consolidate through mergers and takeovers. This development may, in turn, lead food processors to consolidate to meet the large-scale needs of grocery retail chains. Some large wholesalers—concerned about the ability of the smaller, independent food retailers that they supply to compete with retail chains and stay in business—are vertically integrating into retailing by acquiring stores of their own.

Primary differentiation strategies include expanded project offerings—new food product introductions continue to set records, outpacing nonfood packaged goods—along with updated store designs and technologies to improve service. Food companies are also adopting some less conventional methods, focusing on new ways of image enhancement—for instance, publicizing their initiatives to advance social agendas beyond those required by law. The companies are also using new advertising approaches. A shift from TV advertising to other venues, such as magazines, the Internet, and video games, reflects a move from mass to individualized marketing. This trend is also reflected in new product labels designed to appeal to consumer self-image (for instance, with "upscale" terms such as *premium* and *gourmet).*

How Was the Study Conducted?

This study examined recent major developments in the U.S. food marketing system, using data from the U.S. Census Bureau, Bureau of Labor Statistics, Bureau of Economic Analysis, Food Institute, Marketing Intelligence Service's Productscan Online,data base, company annual reports and Web sites, and various trade publications.

INTRODUCTION

The U.S. food marketing system comprises five broad stages of economic activity: production, processing and manufacturing, wholesaling, retailing, and consumption (see box, "Economic Stages of the U.S. Food Marketing System"). The food manufacturing and distribution stages serve as the bridge between production and consumption, coordinating the delivery of farm products in the form, place, and time preferred by consumers. Vertical coordination between stages of the system is achieved through a variety of methods, including spot markets, contracts, alliances, and vertical integration, varying by specific food industry. In 2001, the food and fiber marketing system contributed 12.3 percent to the U.S. gross domestic product and employed 23.7 million people, or 16.7 percent of total U.S. employment. According to calculations based on U.S. Census Bureau data, the food marketing system also accounted for 9 percent of the value of all merchandise exported by the United States in 2005. The activities and services provided by food manufacturers and distributors accounted for about 81 percent of consumer food expenditures in 2002, while the farm value component accounted for the remainder.

The objective of this report is to assess recent developments in the U.S. food marketing system. The report covers many of the same topics as an earlier Economic Research Service (ERS) report, *The U.S. Food Marketing System, 2002* (Harris et al., 2002). However, unlike the previous report, the current report is organized by topic area rather than by stages of the marketing system. In addition, it focuses more on key developments, using the most recent data available.

The report is organized into three sections: (1) Changes in food distribution channels, including the growing diversity of retail outlets selling food and the globalization of food markets—the influx of retailers not traditionally involved in selling food is one of the most important developments in the U.S. food system; (2) changes in market structure related to consolidation, mergers and acquisitions, and methods of vertical coordination; and (3) differentiation strategies in relation to the corporate social responsibility movement, new product introductions, and advertising that has the capacity to influence food consumption behavior and nutrition.

ECONOMIC STAGES OF THE U.S. FOOD MARKETING SYSTEM

The U.S. food marketing system is generally composed of five broad stages. Over 25,000 food and beverage processors purchase output from more than 2.1 million farms. Domestic and foreign food processors handle over 90 percent of the value of U.S. farm production, with the remainder reaching consumers in unprocessed form (Connor and Schiek, 1997).

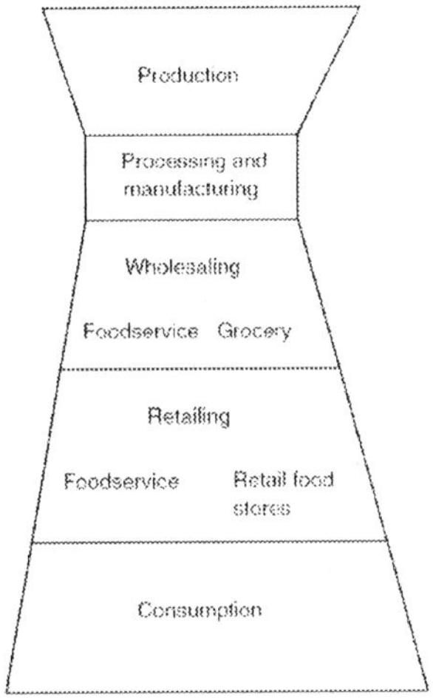

Processed and packaged products are then sold to over 32,559 wholesalers, 112,662 food and beverage retailers, and 377,717 foodservice companies for distribution to over 111 million households. Grocery wholesalers deliver products to retail food stores, such as supermarkets, whereas specialized foodservice wholesalers distribute to foodservice outlets. The foodservice and grocery retail segments compete for the consumer's not compete directly with each other. Foodservice distributors serve more locations with smaller order sizes, as the number of items typically required by a foodservice establishment is considerably smaller than the number offered through grocery stores.

Sources: U.S. Census Bureau (2004, 2005); National Agricultural Statistics Service, USDA (2006).

Competition from Nonfood Stores and Responses by Food Retailers

As food companies strive to grow or maintain market share in a slowly growing domestic food economy, distribution channels for marketing food products in the United States are changing. Over the past 10 years, the food industry has seen an influx of companies not traditionally involved in food sales, led by Wal-Mart supercenters, along with continued growth in the foodservice segment. In addition, global food marketing channels continue to expand through processed food exports and imports and foreign direct investment.

We begin by discussing various strategies of companies as they seek a competitive advantage within the food industry.

Strategies Companies Use for Positioning Themselves

In order to compete, a firm must establish a sustainable competitive advantage. In his theoretical framework for explaining competitive advantage, Porter (1990) posits two basic types of advantage: lower costs and differentiation. Lower costs result from a firm's ability to produce and market products more efficiently than the competition. Differentiation refers to the ability of firms to provide consumers with unique and superior products and service at a premium price. While it is possible for companies to follow both strategies, this is complicated by the additional costs of providing unique quality and services. Nonetheless, companies must be attentive to both types of advantage, while emphasizing one. For example, a company that produces high-quality products but neglects costs can see its price premium offset by higher expenses.

Another important choice is the company's competitive scope, or breadth of the company's target within an industry. A company must choose the products to produce, channels to market the products, types of customers and geographic areas to target, and the industries in which to compete. Segments within an industry have particular needs that require different strategies and capabilities. For example, regular milk and more expensive organic milk are sold to consumers with different tastes and preferences. Companies in the dairy industry may choose different competitive scopes, with some producing both organic and regular milk, for example, while others specialize in organic milk.

Companies may gain competitive advantage by expanding globally or by competing in related industries. For example, Pepsi-Cola and Coca-Cola relied on their extensive bottling and distribution channels and brand-building capabilities to gain important advantages in bottled water production (see "Structural Changes in Food and Beverage Processing," p. 21). A global strategy is one in which a firm sells its product in many nations, using an integrated worldwide approach.

To gain competitive advantage, a company must choose the type of advantage it wishes to achieve and the scope for obtaining it. For conceptualizing, combinations of advantage type and scope form four basic strategies (figure1). For example, companies that choose a broad scope may follow a cost-leader strategy by offering a variety of products at lower prices and less-than-superior quality. Alternatively, companies may follow a differentiation strategy by offering a variety of products at premium prices. Firms choosing a narrow target may follow either a cost-focus or focused-differentiation strategy. Under a cost-focus strategy, companies offer basic products at lower prices than those following a cost-leadership strategy. Finally, the focused-differentiation strategy involves specialized products that command premium prices.

Although industry structure limits the range of approaches, different strategies can successfully coexist in many industries. There may also be variations in how to focus or differentiate within the same strategy. Changes in industry structure, or developments that establish new bases for competing, offer firms opportunities to substantially shift their competitive position.

A "stuck in the middle" strategy refers to companies that follow all positioning strategies simultaneously. These companies are destined to fail because the incompatible nature of the different strategies will result in below-average performance (Porter, 1990).

Strategic Positioning by Nontraditional Food Outlets

By segmenting consumers based on their preferences, nontraditional food retail outlets, including supercenters, warehouse clubs, drugstores, and dollar stores, position themselves through pricing strategy and product variety. Supercenters follow a cost-leadership strategy, offering a wide variety of food and nonfood merchandise at lower prices than traditional stores such as supermarket chains. By competing in both the food and drug industries, drugstores follow a differentiation strategy. Dollar stores have also emerged as formidable competitors to traditional food retailers by appealing to bargain and low-income shoppers with a cost-focus strategy.

Growth of Nontraditional Grocery Retailers

Nontraditional outlets increased their share of U.S. food-at-home expenditures (i.e., food purchased to be prepared and consumed at home) from 17.1 percent in 1994 to 31.6 percent in 2005 (table 1). Most of this growth was due to supercenters and warehouse club stores, which together took in 17.1 percent of food-at-home expenditures in 2005, up from 3.9 percent in 1994. Warehouse clubs compete by catering to small businesses and middle- to upper-income consumers with a limited variety of products and a grocery section dedicated to large-size packages and bulk sales (Brady and Cavanaugh, 2005; Hamstra, 2005).

Other types of nontraditional food outlets have recently emerged as competitors in the food industry (table 2). Drugstores, such as Walgreens and CVS, accounted for 4.8 percent of all food and nonfood grocery sales in 2005, although drug sales remain their core business (Veiders, 2005). Food sales by drugstores reached $7.25 billion in 2004, up 36 percent from 5 years earlier (Alexander, 2004). Each store offers a small assortment of food and beverages, accounting for approximately 5 percent of sales, but with the large number of stores operated by each company (e.g., CVS with 6,100 stores, Walgreens with 5,251, and Rite Aid with 3,323), this adds up to significant food sales. Further, food and beverage sales earn higher margins for the company than drug sales. Customers are attracted by the pharmacy and the convenience of food product offerings.

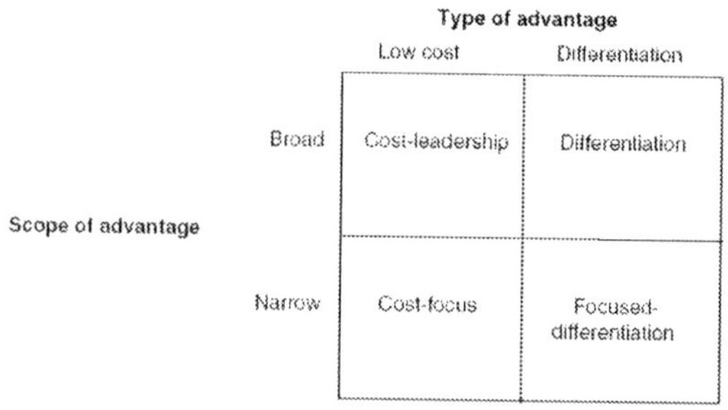

Source: Porter, 1990.

Figure 1. Types of competitive strategies

Table 1. Share of Food-at-home Expenditures by Type of Outlet[1]

	1994	2001	2005
	Percent		
Traditional grocery retailers			
Supermarkets	59.2	62.7	58.2
Convenience	3.1	2.9	2.9
Other grocery	16.6	3.5	3.6
Specialty food stores	2.8	2.3	2.7
Total traditional	81.7	71.4	67.4
Nontraditional grocery retailers			
Supercenters (e.g., Wal-Mart, Super Target, Super Kmart, Meijer, Fred Meyer) and warehouse clubs (e.g., Costco, Sam's, BJ's)	3.9	11.7	17.1
Mass merchandisers (e.g., traditional Wal-Mart, Target, and Kmart stores)	1.8	2.2	1.8
Other stores (e.g., Walgreens, Dollar General)	9.0	9.6	8.7
Home-delivered and mail order	2.4	4.1	4.0
Total nontraditional	**17.1**	**27.6**	**31.6**

[1] Excludes sales by farmers, processors, wholesalers, and others, which have accounted for approximately 1 percent of food-at-home expenditures.

Source: ERS, http://www.ers.usda.gov/ Briefing/CPIFoodAndExpenditures/ Data/ table 16.htm, accessed July 13, 2006.

Table 2. Share of Grocery Sales by Nontraditional Outlets[1]

	1994	2001	2005	2010[2]
	Percent			
Warehouse club	4.8	8.5	7.1	7.3
Supercenters	2	9.9	13.6	20.7
Dollar stores (e.g., Family Dollar, Dollar General, Dollar Tree, Fred's, 99 Cents Only)	na	na	1.7	2.1
Drugstores (e.g., Walgreens, CVS, Rite Aid)	na	na	4.8	4.5
Mass merchandisers	na	na	5.7	5.4
Internet (e.g., FreshDirect)	na	.12	na	na
Military commissaries	na	na	.5	.5
Other (mini-club, deep discount drugstore)	2	.60	na	na
Total nontraditional	8.9	19.14	33.4	40.5

na=Not available.

[1] Sales include food and nonfood grocery items, health and beauty items, greeting cards and magazines, alcohol, and tobacco.

[2] Forecast.

Sources: Caffarini and Cavanaugh, 2006; Griffith, 2002; Rogers, 2000.

Dollar stores are driving sales growth by opening new stores and adding grocery items.[1] Products are often priced at $1 or $2, with grocery products accounting for between 20 and 80 percent of total sales (Brady and Cavanaugh, 2005).

Competition for food sales is widening to include some of the Nation's largest retail companies. In 2003, Sears, Roebuck and Co. introduced its Sears Grand stores, which sell clothing, appliances, tools, and a limited array of grocery products such as dairy, lunch meats, and frozen food. Sears Grand, with large freestanding buildings close to residential areas, emphasizes convenience over selection. A "dollar section" of packaged foods was also added. Sears purchased stores from Kmart, which had recently emerged from bankruptcy, and converted some to Sears Grand stores. In 2005, Sears merged with Kmart to become the Nation's third-largest retailer, referred to as Sears Holdings Corporation. Home Depot, the second-largest U.S. retailer behind Wal-Mart, is testing convenience stores with gas stations adjacent to Home Depot stores.

Wal-Mart's Influence

Supercenters have led the growth in grocery sales by nontraditional outlets (table 2). Primarily through its supercenters, Wal-Mart has emerged as a prominent player in the food industry (Stewart and Martinez, 2002). Within 12 years after opening its first supercenter in 1988, Wal-Mart became the Nation's leading grocery retailer (Appendix A, Appendix table 1). By offering food at lower prices with very low profit, Wal-Mart can attract customers who also buy the store's more profitable general merchandise.

Wal-Mart's tremendous growth is largely due to efficiencies in managing its supply chain to lower costs (Irwin and Clark, 2006). Through an innovative logistics system and collaboration and data-sharing with suppliers, Wal-Mart streamlines the flow of products to consumers and tailors products to individual stores, based on customer preferences in the community where that store operates (Zwiebach, October 2005). Wal-Mart requires manufacturers to manage their own store inventories and makes available its private, invitation-only "Retail Link" system to exchange information with its suppliers over the Internet in a secure environment (Sparks and Wagner, 2003). About 30,000 of its suppliers are using Retail Link to track how well their products are selling by region or store, to review inventory levels, and to assist Wal-Mart in optimizing its inventory holdings. Many of Wal-Mart's products are not stored at the warehouse, but are moved from supplier truck to store-delivery truck through a process called cross-docking.

Wal-Mart has been the prime mover of radio frequency identification (RFID) technology in the United States (Martinez and Stewart, 2003). Items are tagged with an electronic product code (EPC), the digital enhancement of the bar code, containing detailed product data, manufacturing date, expiration and tracking data, and other information. Instead of manually scanning a bar code, an EPC reader automatically transfers the data to a database using radio waves. In 2004, the company introduced the technology in some of its stores and distribution centers, focusing on tagged pallets and cases of higher priced, fast-moving merchandise (Covert, 2005; Sullivan, 2005). By 2007, the company intends to expand its RFID rollout from 500 to 1,000 stores and to double the number of participating suppliers to 600.

The RFID technology has the potential to move products from producer to consumer more quickly, which is especially important for perishable products like fresh fruits and vegetables (Parks, 2005). The technology can reduce instances of "out-of-stock." A study of the effect of RFID on stores by researchers at the University of Arkansas found a 32-percent annual reduction in out-of-stock merchandise for products selling at a rate of 0.1 to 2 units a day (Collins, 2006). The research data also showed a 62-percent reduction for goods selling at a rate of 6 to 15 units a day. For slower and faster moving items, no improvement in product availability was found.

The share of grocery and tobacco sales at Wal-Mart's discount stores and supercenters has been growing steadily, approaching that of the company's own Sam's Clubs, the second-largest wholesale club behind Costco (figure2). In addition, while many food retailers achieve growth through acquisitions, Wal-Mart expands by opening new stores (table 3). Wal-Mart continues to open new, smaller supermarket stores, referred to as "Neighborhood Markets," close to its supercenters to serve as a fill-in stop between customers' trips to the supercenters. The company is also expanding its offerings of private label food items.[2]

Given its immense and growing size, Wal-Mart is changing the traditional relationship between manufacturer and retailer (Greenhouse, 2004). The company is the biggest customer of many of the Nation's leading food processors, and its share is slowly growing (table 4). Regarded as a tough negotiator, Wal-Mart is said to be capable of extracting concessions from suppliers (Connor and Schiek, 1997). As manufacturers' revenue becomes more dependent on Wal-Mart, they are willing to invest in technology to increase efficiency and to satisfy Wal-Mart's requests.[3] To do business with Wal-Mart, manufacturers must comply with company mandates, such as tagging cases with RFID tags and delivering case-ready meats that are prepackaged and priced (Martinez and Stewart, 2003).[4] As manufacturers commit their resources to serving Wal-Mart, they may provide less support for retailers that are unable to provide the same sales growth (Zwiebach, April 2004).[5]

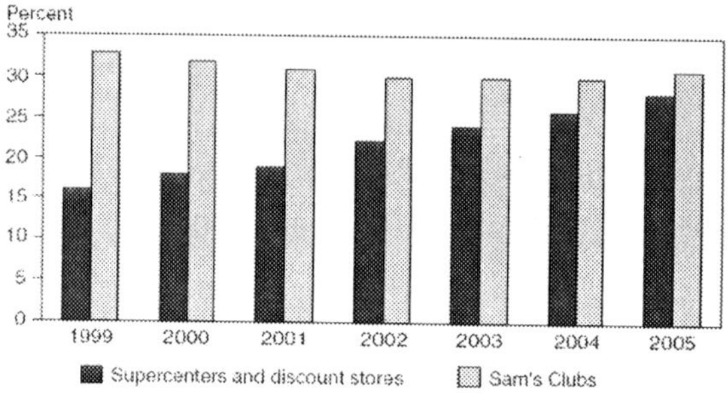

Source: Annual company reports submitted to the Securities and Exchange Commission.

Figure 2. Share of sales by Wal-Mart's supercenter and discount stores and Sam's Club for grocery and tobacco products

Table 3. Wal-Mart Growth

	Unit	1995	2005
Number of distribution centers[1]	Number	30	99
Total number of U.S. stores:	Number	672	2,349
Supercenters[2]	Number	239	1,713
Sam's Clubs[3]	Number	433	551
Neighborhood markets[4]	Number	0	85
New store openings in the United States[5]	Number	97	435
International stores	Number	276	1,587
Total sales	Bil. dol.	82.5	285
Percent of U.S. GDP	Percent	1.32	2.43
Total U.S. employees	Million	0.7	1.3
Percent of U.S. employment	Percent	0.59	0.99

[1] In 2005, approximately 19 percent of Wal-Mart's merchandise was shipped directly to stores from suppliers, up from 16 percent in 1995. The remainder was shipped from Wal-Mart's distribution centers, of which 34 are grocery distribution centers.

[2] Nearly 2,200 Supercenters are expected to be operating by the end of 2006 (Zwiebach, October 2005).

[3] These are Wal-Mart's warehouse club stores.

[4] These are Wal-Mart's supermarket stores.

[5] Includes discount stores relocated or expanded to supercenters.

Sources: Annual reports filed with the Securities and Exchange Commission.

Foodservice Gains

Grocery retailers are also facing formidable competition from the foodservice channel as consumers slowly increase their share of food expenditures from foodservice outlets. In 2005, the food-away-from-home market captured 48.5 percent of total food expenditures, compared with 46.1 percent in 1994 (U.S. Department of Agriculture, Economic Research Service). To continue building market share, foodservice companies are following a differentiation strategy, offering new products and services that cater to the lifestyles of today's health-conscious, time-pressed, and demanding consumers.

In addition to their traditional food offerings, foodservice companies have responded to consumers' diet and health preferences with a host of new products. Facilitated by advances in fresh-cut produce technology, McDonald's introduced sliced apples as a substitute for fries in Happy Meals in 2004. The Nation's largest fast food company procured more than 54 million pounds of fresh-cut apples, becoming the largest consumer of apples in the foodservice industry (Enis, 2005). More recently, the company introduced an apple, grape, and walnut salad. Mineral coatings and modified-atmosphere packaging are extending the shelf-life of fresh-cut items, allowing a wider variety of offerings. New chicken products were recently launched by several of the leading fast food chains, among them McDonald's Premium Chicken Sandwiches and Chicken Selects, Burger King's Chicken TenderCrisp sandwich, Wendy's Chicken Temptations sandwiches, and KFC's new roasted chicken menu. In response to consumer demand for low-carbohydrate foods, full-service chains, such as Applebee's and TGI Friday's, have developed low-carb menus, while fast food chains, such as Hardee's and McDonald's, offer burgers without the bun.

Companies continue to offer consumers a variety of large-portion items that are high in fat and calories. For example, in 2004, Hardee's introduced a 1,420-calorie "Monster Thickburger." A nutritional calculator on Hardee's Web site allows consumers to analyze the nutritional content of all its menu items, including calories, calories from fat, and sodium.

McDonald's is providing consumers with on-the-spot nutritional and product-sourcing information. In February 2006, the company introduced product packaging that displays the nutritional content of most of its products. Five icons provide information about the calorie, protein, fat, carbohydrate, and sodium content compared with daily nutrient recommendations (Muirhead, 2005). By the end of 2006, the company expected to have the nutrition labeling in 20,000 of its 30,000 restaurants. It also added to its Web site an "open door" tour with information on how its products are produced, beginning at the farm.

Some limited-service chains are following a focused-differentiation strategy by catering to consumer preferences for "casual indulgence" with more "upscale" items. Some of the fastest growing chains are those offering specialty products and gourmet ingredients. From 2004 to 2005, Starbucks, the world's leading retailer of specialty coffee, had the third-largest percentage increase in sales among limited-service chains and the second-largest increase in units (Technomic Information Services, Inc., 2006). The company now ranks as the sixth-largest foodservice chain (Appendix A, Appendix table 1). Panera Bread and Quiznos Sub also ranked among the 10 fastest growing chains in terms of both sales and units. Panera Bread sells all-natural specialty breads, while Quiznos Sub is an upscale chain serving sandwiches made from proprietary bread and special ingredients, such as Tuscan basil mayonnaise and spring-mix lettuce. Mainstream chains are following suit with their own unique product offerings. Through a partnership with a coffee processor, McDonald's is offering "Newman's Own Organics Blend" of Fair Trade Certified, organic specialty coffees (American Institute of Food Distribution, Inc., October 31, 2005). Chick-Fil-A recently introduced its "Café Blends Coffee" premium line.

Table 4. Percent of Manufacturer Revenue from Sales to Wal-Mart, 2002-2004

Manufacturer	2002	2003	2004
	Percent		
Procter and Gamble	17	18	17
Dean Foods[1]	10.6	13.4	14.6
General Mills	12	13	14
Kellogg Company	12	13	14
Kraft Foods	12.2	12	14
Campbell Soup	12	12	13
Tyson Foods[2]	Less than 10	Less than 10	12
Pepsico[3]	Less than 10	10	11

[1] Dairy division. In 2002, the second-leading customer accounted for 7.5 percent of sales.

[2] According to Tyson's 2004 annual report filed with the SEC, "Any extended discontinuance of sales to this customer could, if not replaced, have a material impact on the Company's operations; however, the Company does not anticipate any such occurrences due to the demand for its products."

[3] In all of North America, Wal-Mart accounted for 14 percent of Pepsico's sales revenue in 2004, up from 12 percent in 2003.

Sources: Annual reports filed with the Securities and Exchange Commission.

Full-service restaurants are differentiating themselves by expanding their takeout business (Technomic, Inc., 2005). In 2004, their takeout sales were estimated to be nearly 10 percent of total sales. Over the previous 3 years, takeout sales in full-service restaurants grew about 8 percent per year, roughly twice the rate of their total sales. Many of these restaurants have added reserved parking spaces and special entrances for takeout orders, and some, such as Outback Steakhouse, are adding drive-up service.

Strategic Responses by Traditional Grocery Outlets

Traditional food retailers that follow a "stuck-in-the-middle" strategy will likely continue to lose market share to nontraditional retail outlets and food-service companies. Traditional outlets that have higher costs than Wal-Mart, fail to differentiate from the competition, and fail to compete in a narrower market segment lack any competitive advantage.

Much has been written in the popular press about conventional supermarkets caught between the "big box" stores (e.g., supercenters) and specialty stores that target a specific consumer segment, such as the "fresh format" stores (Brady and Cavanaugh, 2005). These are stores exceeding 20,000 square feet that follow a focused-differentiation strategy by emphasizing perishables, with ethnic, natural, and organic assortments that differ from those of traditional retailers. From 1999 to 2004, sales at the two largest publicly traded fresh format stores that specialize in organic and natural foods, Whole Foods and Wild Oats, grew by 159 and 45 percent (annual reports filed with the Securities and Exchange Commission), compared with a 13-percent increase for the entire U.S. grocery store industry (U.S. Census Bureau). Over the same period, sales by United Natural Foods, a leading wholesale distributor of organic and natural foods, increased by 95 percent, compared with a 16-percent increase for all merchant food wholesalers, excluding manufacturers' sales branches and offices.

Conventional outlets are responding with a number of competitive strategies to retain market share. Establishing a competitive position requires an understanding of what factors influence costs, which attributes are preferred by consumers, and how cost factors and attributes vary by customer segment (Besanko et al., 1996). Strategies for standing out from the competition, lowering costs, and streamlining product delivery are discussed next.

Differentiation Strategies

Traditional grocery stores are differentiating from the competition with expanded product offerings, new store layouts, and innovative in-store technologies. Mainstream supermarkets accounted for the largest share of the organic foods market in 2004, with 37 percent of total sales, followed by independent natural food stores (28 percent) and natural food supermarket chains (19 percent).[6] Some supermarkets are opening their own organic and natural food stores and producing their own corporate-brand organics. For example, Supervalu recently opened its first Sunflower Market banner store that features organic, minimally processed products with no preservatives. Some of these products are sold as private label under the "Nature's Best" brand. Other companies offering their own corporate-brand organic products include Kroger (Naturally Preferred), Giant Food (Nature's Promise), and Shaw's (Wild Harvest).

Private labels are one way that retailers have diminished the market power of manufacturers, differentiating themselves with store-brand products of better value than national brands (Kinsey, 2001). As Wal-Mart and super-centers increase their private label products, traditional retailers are introducing or expanding their own private label programs. Sales of private label products in supermarkets rose by 0.9 percent in 2004, compared with national brand growth of 0.6 percent. Private label share of supermarket sales is growing slowly, increasing from 14.0 percent in 1994 to 16.1 percent in 2004, but this represents about $10 billion of additional sales. Examples include H.E. Butt's Central Market Organic and All-Natural line of food products, Albertsons' Essencia premium line, Publix's line of branded Hispanic products, and Food Lion's Butcher's Brand Premium Beef. Premium brands project an image of higher quality based on superior food ingredients or packaging (Connor and Schiek, 1997). Food Lion's premium beef, for example, is sourced from Midwestern producers that comply with the company's standards for beef quality, including marbling and aging.

Major food chains and manufacturers are forming innovative partnerships as retailers attempt to differentiate themselves with their own line of products (Forbes.com, 2004). Private label processors include large national brand manufacturers that use their expertise and excess plant capacity to supply store brands; small manufacturers who specialize in particular product lines and produce store brands almost exclusively, often owned by corporations that also produce national brands; and regional brand manufacturers that produce private label products for specific markets (Private Label Manufactures Association, 2006). Large manufacturers, such as McCormick, Del Monte, and

Birds Eye, usually do not cannibalize their national brands with store-brand versions of their own products (ConsumerReports.org, 2005). For example, Del Monte produces soup for retailers, but not fruit products. When companies do make national and store brands of the same product, the products may have different attributes. For example, the quality of farm output in the ingredients may vary by growing conditions and region. In addition, retailers that emphasize costs may prefer a lower priced version and lower quality ingredients.

To benefit from the growing popularity of dollar stores, some grocers are adding more dollar merchandise (i.e., goods priced at one dollar). More than 27 percent of new grocery stores had aisles designated for dollar merchandise in 2004, compared with 6 percent in 2002 (Adamy, 2005). In 2002, Save-A-Lot, a small discount grocery store chain owned by Super-valu, bought a dollar store chain to get more dollar merchandise into its stores, including videos, jewelry, and household goods. Winn-Dixie is also adding more dollar merchandise to improve sales.

As supermarkets compete with the foodservice sector, their annual sales of prepared food are growing at about 4-4.5 percent, compared with 2-2.5 percent for their other food items (Rogers, 2005). For example, they are offering a variety of precooked, precut, or premarinated meats for warming or cooking.

In addition to supercenters and wholesale clubs, mainstream supermarkets such as Kroger and Albertsons are adding fuel pumps to entice motorists to purchase food and other products when they buy gas, and some stores offer discounts on fuel for purchasing promotional items. According to a 2006 Food Marketing Institute survey, a third of member grocers had fuel pumps, compared with 6 percent in 2003 (Showalter, 2006).

New store designs offer consumers an upscale shopping experience. Stop & Shop Supermarkets rents space to Dunkin' Donuts and Boston Market and is redesigning its health and beauty section to emphasize "health and relaxation." Safeway is opening "lifestyle" stores, sophisticated shopping venues with high-quality produce, soft lighting, and classes on topics such as flower arranging (Adamy, 2005). Safeway-owned *Dominick's Lifestyle* stores feature an expanded section of organic foods and an upscale bakery, deli, wine, and floral section. The stores also have a Starbucks coffee shop with sofas and plush chairs. *Kroger Marketplace*, twice as big as a regular Kroger store, features a Starbucks and Donato's Pizza stand. Building on Kroger's acquisition of Fred Meyer, a large supercenter selling groceries and merchandise, the new stores offer furniture, linen, electronics, toys, and household appliances in addition to grocery items. A&P, the 11th-ranked supermarket chain, is expanding its upscale Fresh Market concept, featuring natural and organic produce and restaurant-style hot and chilled foods.

Innovative, timesaving technologies offer another means for traditional grocers to differentiate from the competition. One of the fastest growing technological applications in new grocery stores is self-checkout lanes. In 2005, nearly 56 percent of food retailers used self-checkout systems, up from 38 percent in 2004 (American Institute of Food Distribution, Inc., October 3, 2005). The technology has become so ingrained that it is no longer considered a major initiative by food retailers (Garry, January 31, 2005).

Biometric technology allows customers to pay at checkout by scanning a finger to match the fingerprint on record, which releases payment from a bank account or credit card. According to the 2004 *Supermarket News* annual survey of corporate and information technology, executives at 68 food retail and wholesale companies, representing 36,000 supermarkets, 25 percent of the companies planned to use or test biometric technology in 2005, up from 15 percent in 2004 (Garry, January 31, 2005). Stop and Shop Supermarkets, owned by Ahold, is introducing the "Shopping Buddy" into some of its stores, an updated version of the VideOcarts introduced a dozen years ago (Garry, November 2005). The computerized self-scanning device attaches to the front of a shopping cart and provides a wide range of information, some of it targeted to the individual loyalty-card shopper. Shoppers can scan and bag items in the aisle, locate products, check prices, order deli items without standing in line, and maintain a running total of items in their cart. The screen also informs shoppers of favorite items on sale as they approach those items in the aisle. By swiping their loyalty card at checkout, the cashier can total their purchases without scanning individual items.

Cost-Lowering Strategies

In addition to consolidation through mergers and acquisitions (see section "Structural Changes in Food Distribution"), traditional food retailers are reducing costs through supply chain initiatives and restructuring of operations. They are attempting to improve inventory control and consumer response by emulating Wal-Mart's supply chain management practices (McKinsey Global Institute, 2001).[7] A 1992 industry-wide initiative, known as Efficient Consumer Response (ECR), was an attempt by retail grocers to adopt Business-to-Business (B2B) e-commerce practices similar to Wal-Mart's to streamline inventory ordering and reduce costs. However, the initiative waned due to computer systems that were either nonexistent or incompatible with those of suppliers (Mohtadi and Kinsey, 2005). The effort was further hampered by retailers' reluctance to share sales information with manufacturers, adoption of technologies that proved unprofitable, and perceived unequal sharing of ECR benefits across the supply chain.

ECR, however, paved the way for many efficient management practices and led to a new B2B Internet platform, referred to as UCCnet (Kinsey, 2000). UCCnet is a division of GS1 US, formerly the Uniform Code Council (UCC), which is the organization best known for developing bar codes containing a Universal Product Code (UPC). To increase consistency in product and invoice data used by food distributors and suppliers, UCCnet was the primary driver of data synchronization between retailers and manufacturers in the United States (Seifert, 2002; Garry, March 2005). In 2004, 29 percent of retailers and wholesalers surveyed by *Supermarket News* were involved in data synchronization with trading partners, 29 percent were planning to participate in 2005, and 33 percent were considering it (Garry, January 31, 2005). By allowing retailers and suppliers to electronically exchange information and services in a secure environment, UCCnet facilitates the delivery of products to reduce out-of-stocks and excess store inventory.

A global data system was needed, and in 2004 the Global Data Synchronization Network (GDSN) was introduced. The GDSN allows certified data pools, such as UCCnet, Transora, and WorldWide Retail Exchange (WWRE), to exchange standard product data registered in the GS1 Global Registry.[8] The GS1 Global Registry is a centralized directory that validates and stores basic item and participant information. It also serves as a "pointer" to the appropriate data pool for obtaining information on a specific item or party. From January to June of 2005, the number of trading partners participating increased by 16 percent, and registered items (Global Trade Item Number) increased by 62 percent.[9] In June 2005, there were 10 certified data pools, with 252 manufacturers and 17 retailers synchronizing data (Garry, June 20, 2005). By May 2006, 5,000 trading partners, mostly manufacturers, were using the GDSN (Garry, June 5, 2006).

Food retailers have also moved forward with global scanning standards. To facilitate global trade, the UCC selected January 2005, referred to as "Sunrise 2005," as the deadline for North American retailers to become capable of scanning the 13-digit European Article Number (EAN) bar code and storing the code in their database. A 2005 Food Marketing Institute survey of information technology executives at 28 food retail and wholesale companies, representing 6,724 U.S. supermarkets, found that all retailers were capable of scanning the EAN (Food Marketing Institute, 2005). All of the companies had dedicated funds for complying with the scanning standards.

The 2004 *Supermarket News* annual survey of food retail and wholesale companies found that 21 percent of companies planned to launch or test RFID in 2005, compared with only 3 percent in 2004 (Garry, January 31, 2005). In 2004, Albertsons asked its top 100 suppliers to begin placing RFID tags on cases and pallets. Publix recently initiated an RFID pilot program through a partnership with the University of Florida (Parks, 2005).

Some of the larger chains are restructuring their operations to improve efficiencies and services, cut costs, and maintain competitive prices. Strategies include closing older stores and focusing on core market areas and assets. In 2005, Winn-Dixie, the Nation's 10th-largest grocery retailer, filed for bankruptcy protection to reposition itself for future growth. The company announced a plan to sell a third of its stores so that it can focus on its strongest markets. The company is also closing some of its distribution centers and selling its dairy, pizza, and beverage/condiment manufacturing plants.

In 2005, A&P announced plans for a major strategic restructuring, including initiatives to improve labor productivity and reduce operating and administrative costs. Shortly afterwards, the company moved away from self-distribution by entering into an agreement to transfer its U.S. distribution operations to C&S Wholesale Grocers. This is expected to reduce costs by harnessing C&S's logistics expertise and purchasing efficiency.

Global Competitive Strategies

In a slowly growing domestic market, firms may gain competitive advantage or offset domestic disadvantages by following a global strategy (Porter, 1990). Advances in information, communication, and transportation technology, along with various trade agreements, have increased access to international markets. In addition, globalization of processing and distribution are associated with rising incomes and international travel, which influence the diversity of consumer food preferences. A global strategy involves an integrated approach to selling products abroad, including information on where best to locate and how to coordinate activities across nations to lower costs or differentiate from global competitors.

Processed Food Trade

International trade has been the fastest growing marketing channel for food processors since the 1970s (Connor and Schiek, 1997). In 1991, U.S. exports of high-value products surpassed bulk agricultural product exports (e.g., wheat, corn, and tobacco) (figure 3). High-value products are processed products not yet ready for final consumption (e.g., wheat flour, soybean meal and oil, and animal hides) and those requiring little or no additional processing (e.g., meats, fresh fruits and vegetables, and nuts). Leading products include red meats, fresh and processed fruits and vegetables, poultry meat, and nuts, which together accounted for 39 percent of all high-value exports in 2005. With an export-based global strategy, food processors can pursue growth opportunities in rapidly expanding economies.

U.S. imports of food products provide more food choices for U.S. consumers, but also bring added competition to the U.S. market. From 1995 to 2005, imports of high-value food products grew at an average annual rate of 8.4 percent, compared with a 2.9-percent increase in high-value exports. The United States has operated at a processed-food trade deficit every year since 2002, widening from $3.3 billion in 2002 to nearly $12 billion in 2005. About 26 percent of all U.S. agricultural imports are classified as noncompetitive, including tropical products that cannot be produced profitably in the United States (e.g., coffee and bananas) (FAS/USDA). Imports also include products that compete directly with domestic production, such as wine and beer, red meats, snack foods, and processed fruits and vegetables.

Foreign Direct Investment

Most global strategies integrate trade and foreign direct investment (Porter, 1990). In 2001, food manufacturing companies accounted for over 75 percent, or $155 billion, of foreign sales at U.S. food facilities abroad (figure 4). Sales by U.S.-owned manufacturing plants located in foreign countries have outpaced the value of processed food exports in recent years (figure 5). Advantages of foreign direct investment compared with international trade include lower transportation costs, greater insight into local consumers and policies, and circumvention of tariffs and domestic market restrictions (e.g., environmental, antitrust) (Connor and Schiek, 1997).

Retailers accounted for most sales by foreign food companies located in the United States, followed closely by manufacturers (figure 6). U.S. sales by retail foreign affiliates, including Ahold (Netherlands) and Delhaize Group (Belgium), were 10 times greater than sales by U.S. retailers located abroad. In addition to U.S. imports of processed food, foreign manufacturers located in the United States are a source of competition for U.S. food manufacturers.

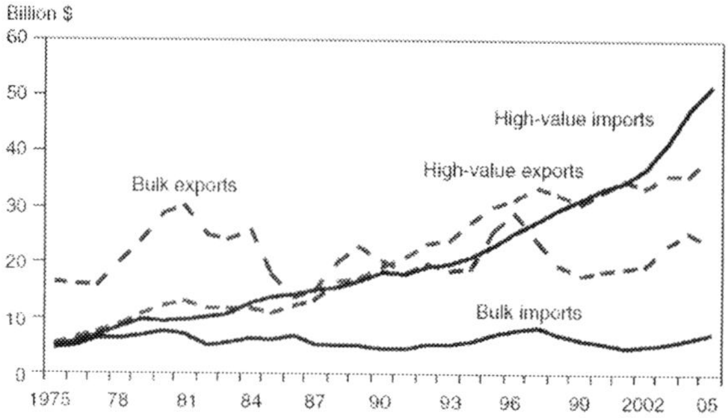

Source: Foreign Agricultural Service, USDA.

Figure 3. U.S. exports and imports, 1975-2005

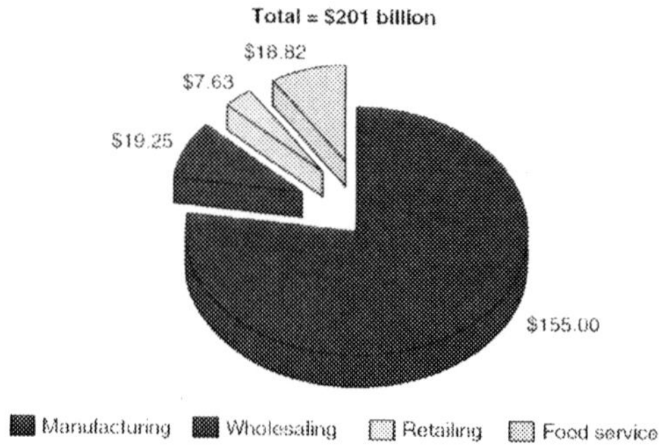

Source: Bureau of Economic Analysis. U.S. Department of Commerce.

Figure 4. Sales by U.S. food firms located in foreign countries ($ billion), 2001

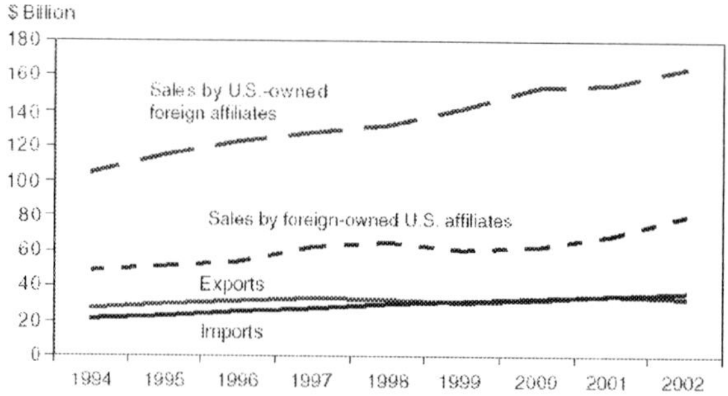

Note: Includes food manufacturing affiliates only.

Source: Bureau of Economic Analysis, U.S. Department of Commmerce.

Figure 5. U.S. processed food exports and imports, and sales by foreign and U.S. affiliates, 1994-2002

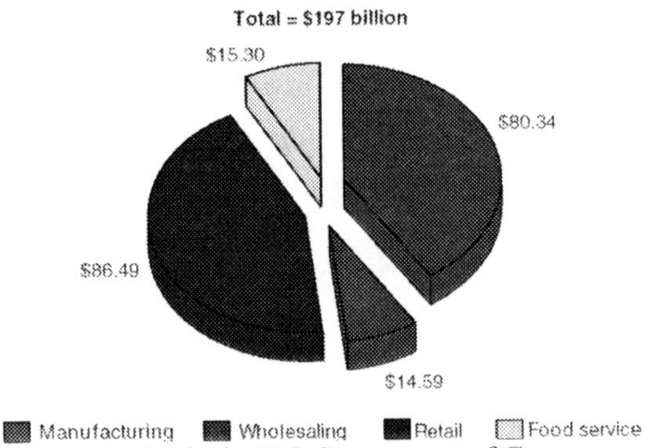

Source: Bureau of Economic Analysis. U.S. Department of Commerce.

Figure 6. Sales by foreign food firms located in the United States ($ billion), 2001

Increases in Food Retail Concentration Resonate Throughout the Food Industry

One widely used measure of market structure is market concentration, or the degree to which economic activity is concentrated in the hands of a few large firms. Many factors contribute to increasing concentration in the U.S. food system. For instance, a firm that follows a lower cost strategy may exploit economies of scale and scope to lower average costs relative to rivals producing smaller quantities of the same product. Economies of scale allow larger volumes to be produced at lower per unit production costs. Economies of scope exist if a firm achieves cost savings as it increases the variety of goods it produces. Economies of scale and scope not only affect the size of firms, but also influence business strategies, such as decisions on merging with another firm or on whether market expansion can be achieved through long-term cost reductions (Besanko et al., 1996). For example, Wal-Mart's demand for low-cost products may influence merger decisions by manufacturers needing to achieve economies of scale and increased efficiency in order to supply Wal-Mart (Hopkins, 2003).

Other factors that can increase concentration include the exit of firms unable to compete with larger, more efficient firms, the cost of new technology, consolidation to offset market power as concentration increases at other stages of the food marketing system, and slow overall demand growth. When demand is falling, new facilities may be discouraged. Consolidation may also be encouraged by the prospect of other economies resulting from size, such as increased access to capital for research and advertising, volume-based price reductions on production inputs, or price premiums for large volumes of specific outputs.

Structural Changes in Food Distribution

Concentration ratios published by the U.S. Census Bureau provide useful summary indicators of the importance of large firms in the food system, as classified under the six-digit North American Industrial Classification System (NAICS) (Appendix B). In grocery retailing, consolidation led to sharp increases in national concentration from 1997 to 2002 (figure7).[10, 11] Mergers and acquisitions can have profound effects on the number and size of companies. In 1998 and 1999, some notable supermarket mergers and acquisitions occurred (Bjornson and Sykuta, 2002; Harris et al., 2002). They included:

- The $12.75-billion acquisition by Kroger of sixth-ranked Fred Meyer, which had previously acquired Smith's Food & Drug Centers and Quality Food Centers.
- Fourth-ranked Albertsons' $12-billion acquisition of second-ranked American Stores Company.
- Ahold's purchase of Giant Food, Inc.

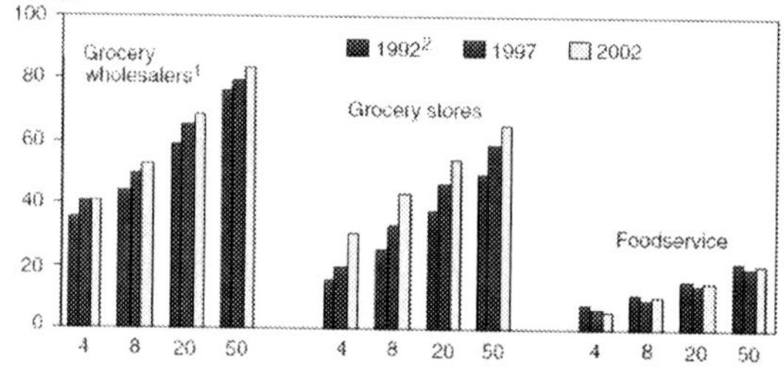

[1] General-line grocery merchant wholesalers, excluding specialty and limited-line merchant wholesalers and manufacturers' sales branches and offices.
[2] Based on Standard Industrial Classification (SIC) system.
Source: Census Bureau, U.S. Department of Commerce.

Figure 7. Percent of U.S. grocery wholesale, retail, and foodservice sales by the 4, 8, 20, and 50 largest food companies

Safeway made several smaller acquisitions, including Vons Companies in 1997, Dominick's Supermarkets in 1998, Randall's Food Markets and Carr-Gottstein Foods Company in 1999, and Genuardi's Family Markets in 2001 (Bjornson and Sykuta, 2002; Sharkey and Stiegert, 2006).

A number of forces heightened competition among larger traditional grocery retailers and led to consolidation through mergers to capture economies of scale (Kaufman, 2000). These forces included increased consumer spending for prepared foods and meals away from home and the growth of food sales by nontraditional competitors such as Wal-Mart. Bureau of Labor Statistics data indicate that labor productivity (i.e., output per worker) in grocery stores increased from 1997 to 2002, following general declines after the mid-1980s (table 5). This may reflect consolidation in the retail grocery sector, more rapid adoption of new technology, and other cost-cutting moves in response to strong price competition from the nontraditional outlets. As discussed earlier, nontraditional outlets are important because of the efficiencies they can trigger in other retailers, in addition to the options they provide consumers. Leibtag (2006) found that the entry and expansion of lower priced nontraditional stores led to further food expenditure savings because of the increased competition that they generate.

Past studies have examined the effect of mergers and acquisitions on operating efficiencies and food prices. Bjornson and Sykuta (2002) analyzed the effect of acquisitions by Albertsons, Kroger, and Safeway on their financial performance over the 1993-1999 period. They found improved cost control and buying efficiencies for Kroger and Safeway, while Albertsons showed signs of inefficiencies from its acquisitions. They also found that gross margin increased for each firm, but they were unable to distinguish how much of the increase was due to greater market power or to improved operating efficiencies and product offerings. A more recent study by Sharkey and Stiegert (2006) examined the effect of supercenters and retail consolidation on food prices in major metropolitan areas from 1993 to 2003. They found that if supermarkets did experience efficiency gains from the mergers, cost savings were not passed on to consumers.

As table 5 shows, a decline in grocery stores from 1997 to 2002 was accompanied by stagnant growth in real grocery store sales, relative to the foodser-vice sector. An expanding foodservice sector and continual growth in the number of foodservice companies and establishments have kept concentration levels in check.[12] Small foodservice establishments can stay competitive by differentiating themselves by their menu offerings and service (Connor and Schiek, 1997).

Table 5. Number of Food Establishments and Companies, Real Sales, and Worker Output, 1992, 1997, and 2002

	Grocery wholesalers[1]		Grocery stores		Foodservice	
Number of establishments:						
1992[2]	3,740		104,105		433,608	
1997	3,390		96,542		486,906	
2002	3,212		95,514		504,430	
Number of companies:						
1992[2]	3,133		72,274		331,488	
1997	2,802		6 8,006		365,588	
2002	2,812		67,757		377,717	
Real sales (percent change):						
1992[2] to 1997	-10.1		-3.9		15.6	
1997 to 2002	15.0		1.7		12.5	
Output per worker (percent change per year[3]):	Food	All[4]	Food	All[5]	Food	All[6]
1992	3.5	3.4	-0.8	1.6	-0.2	na
1997	-0.6	3.2	-1.5	3.2	-0.1	na
2002	0.6	4.4	1.0	4.6	0.8	na

na=Not available.

[1] General-line merchant wholesalers, excluding specialty and limited-line merchant wholesalers and manufacturers' sales branches and offices.

[2] Census year 1992 is based on the Standard Industrial Classification (SIC) system.

[3] Average annual change since the previous Census year.

[4] All wholesaling.

[5] All retailing.

[6] Not available for all accommodations and food services.

Sources: Compiled by ERS from data from Bureau of Labor Statistics and Census Bureau, U.S. Department of Commerce.

Rising concentration in the grocery retail segment is of concern to manufacturers because it makes them dependent on fewer retailers for maintaining their market share. Growing retail concentration could indicate a shift in bargaining power from manufacturers to grocery retailers (Kaufman, 2000; Connor and Schiek, 1997). If food processors must sell their products to a few large distributors, the bargaining power of distributors may increase.

Manufacturers and large retailers offering a broad assortment of items find it advantageous to negotiate directly, reducing the power and influence of traditional wholesalers. Self-distribution is the preferred method of coordination for large grocery chain stores. In 2001, 81.9 percent of chain stores (with "chain" defined by the grocery industry as 11 or more stores under the same ownership) were self-distributing (American Institute of Food Distribution, Inc., Food Institute Report, February 3, 2003).[13, 14]

Structural Changes in Food and Beverage Processing

Concentration at the food manufacturing stage is of special concern to farmers because manufacturers are the primary purchasers of agricultural products. Concentration of ownership has been steadily increasing. According to the Census Bureau, the top 50 processors accounted for 53 percent of food processing sales in 2002, up 15 percentage points, or 39 percent, since 1972 (table 6).

Every manufacturing establishment, or plant, is owned by at least one company. Large companies usually own several operating establishments and often operate plants in several industries (Connor and Schiek, 1997). If ownership is consolidating, the number of companies will fall faster than the number of plants. When firm divestitures outweigh mergers, the number of companies could increase, even though the number of plants is steady or falling.

An increase in the number of manufacturing plants and companies from 1997 to 2002 continues a trend dating back to 1987 (table 6). According to analysis of plant entry patterns by Connor and Schiek, most new production since the mid-1980s stems from the entry of small, startup plants typically owned by small new companies in rapidly growing niche markets. Small plants accounted for most of the net increase in food manufacturing plants in 2002, but their share of shipment value remained below 5 percent (table 7). Many of the smallest plants are open for only a few weeks of the year, often operating with unpaid family labor.

In general, consolidation and investment in optimum-sized processing plants are expected to lead to productivity increases, assuming that the acquiring firms are more efficient operators.[15] Annual average increases in food processing productivity from 1997 to 2002, as measured by labor productivity (output per worker), reached its highest level since the early 1980s. However, since the early 1980s, there has been a general reduction in the growth of labor productivity and multifactor productivity, which compares output growth to changes in all input requirements. Connor and Schiek (1997) offer several possible explanations for this finding. First, the upward trend in new products (see section, "New Product Introductions," p.33) may require shorter production runs and more time for factory line changeover, and may also increase the emphasis on new products at the expense of new process development. Second, an increase in quality control and quality assurance programs may also have led to additional costs and inputs. Third, to the extent that quality improvements lead to higher product prices, price deflators in productivity calculations will overstate purely inflationary changes because quality improvements are not recognized, leading to an underestimate of the change in real output and productivity.

Specific Food Processing Industries

Consolidation can vary widely across food and beverage manufacturing industries, as defined by the NAICS industry classifications. However, the NAICS system is less than ideal for measuring concentration because some industries are defined too narrowly, while other definitions are overly broad. For example, products from the beet sugar and cane sugar refining industries are near-perfect substitutes, which suggests that combining these industries for analysis would provide a more accurate assessment of concentration. On the other hand, the animal slaughtering industry provides an imperfect measure of concentration because output from cattle and hog slaughter plants is combined. Despite these weaknesses, Census industry classifications are the only publicly available source of food processing sales concentration.[16]

Table 6. Number of Food Processing Plants and Companies, Sales Concentration, and Productivity Changes, 1963-2002

Census year[1]	Plants[2]	Companies[2]	Top 50 share of Shipment value	Outp ut per worker		Multifactor productivity[3]	
				Food	All[4]	Food	All[4]
	Number	*Number*	*Percent*	*Percent per year*[5]			
1963	37,521	32,617	32	2.8	2.7	1.4	3.0
1967	32,517	26,549	35	3.2	2.9	0.8	1.5
1972	28,193	22,172	38	3.1	3.6	1.2	1.9
1977	26,656	20,616	40	3.2	3.0	-1.1	-0.3
1982	22,130	16,813	43	3.2	1.5	2.1	0.2
1987	20,624	15,692	47	2.0	3.5	-0.2	2.4
1992	20,798	16,075	50	1.1	2.7	-0.3	0.1
1997	21,805	17,221	51	0.9	3.8	0.0	1.9
2002	23,338	18,696	53	2.5	3.7	-1.0	1.7

[1] Based on Standard Industrial Classification (SIC) system prior to 1997.

[2] Excludes retail bakeries, which were included under NAICS in 1997 and 2002.
[3] For 2002, average change in multifactor productivity index is from 1997 to 2001, the last year available.
[4] All manufacturing.
[5] Average annual change since the previous Census year.
Sources: Compiled by ERS from Connor and Schiek (1997), Bureau of Labor Statistics, and Census Bureau, U.S. Department of Commerce.

Table 7. Size Distribution of Food Processing Plants, 1997 and 2002

Size (employees)	1997				2002			
	Plants		Value of shipments		Plants		Value of shipments	
	Number	*Percent*	*Million $*	*Percent*	*Number*	*Percent*	*Million $*	*Percent*
1-19	17,576	66.8	17,143	4.1	19,179	68.7	19,973	4.4
20-99	5,509	21.0	81,237	19.2	5,478	19.6	83,665	18.2
100 or more	3,217	12.2	323,988	76.7	3,258	11.73	55,149	77.4
Total	26,302	100.0	422,368	100.0	27,915	100.0	458,787	100.0

Source: Compiled by ERS from data from U.S. Census Bureau, 2002 and 1997 Economic Census, Manufacturing, General Summary.

In 2002, the latest Census year, the average national market 4-firm concentration ratio (i.e., CR4—the percent of shipment value accounted for by the four largest firms) across 44 food and beverage manufacturing industries was 50.2 percent, up only slightly from 49.8 in 1997.[17] Five industries had relatively low levels of national concentration (CR4 less than 30), while eight had high levels (CR4 greater than 70) (table 8). The soybean processing industry was the only highly concentrated industry in 2002, with a net increase in plants between 1997 and 2002 (Appendix A, Appendix table 2). Soybean processing is particularly important to agriculture, providing a major feed ingredient, product exports, and food uses such as cooking oil (Ollinger et al., 2005). The increase in soybean processing plants in 2002, and the decline in plant size (output per plant) and worker productivity, run counter to long-term trends from 1977 to 1997 (table 9).

Most of the additional soybean processing plants were in the small-size category (1-19 employees), which accounted for a slightly larger share of shipment value, from 1.6 to 2.3 percent. According to plant entry and exit patterns from 1972 to 1992, plant size of oilseed (including soybean, wet-corn milling, and cottonseed) processing entrants was about 50 percent of the industry mean size, and 60 percent of these plants exited within 5 years (Ollinger et al., 2005).

Table 8. National 4-firm Concentration Ratios[1] (CR4) for Food and Beverage Manufacturers, 2002

CR4<30 (Low)[2]		CR4>70 (High)[3]	
Industry	**CR4**	**Industry**	**CR4**
Fresh and frozen seafood processing	20	Breweries	91
Meat processed from carcasses	24	Malt manufacturing	91
Perishable prepared food	24	Beet sugar	80
Fruit and vegetable canning	24	Soybean processing	80
Spice and extract manufacturing	29	Breakfast cereal	78
		Other snack food[4]	75
		Distilleries	71
		Specialty canning	71

[1] CR4 = The percent of shipment value accounted for by the four largest firms.

[2] Excludes animal feed and retail bakeries, which are local market industries, and "all other miscellaneous food." CR4 is meaningless for the miscellaneous food product category.

[3] In 2002, concentration was not provided for cane sugar refining and flavoring syrup and concentrate to avoid disclosing individual company data. In 1997, CR4 for these industries was 99 and 81, respectively.

[4] Excludes roasted nuts and peanut butter.

Source: U.S. Census Bureau, U.S. Department of Commerce.

Over the 1997-2002 period, nearly two-thirds of food manufacturing industries became more concentrated (Appendix A, Appendix table 2). Eighteen had sizeable increases in concentration, each exceeding 10 percent (table 10). Industries with processing operations that serve local markets, with the exception of retail bakeries, also had fewer firms. Firms serving local markets have historically exited at a faster rate than other food processors, which suggests that merger activity has been more intense in these industries (Connor and Schiek, 1997).

Falling demand (i.e., reduction in sales volume) was associated with the disappearance of plants and firms in seven of the industries, including flour milling and fluid milk (table 10). In flour milling, increases in average plant size, worker productivity, and concentration were consistent with historical trends (table 9). Increases in concentration beginning in the early 1970s reflect the growing presence of multiplant grain companies, such as Cargill (Ollinger et al., 2005). However, declining per capita flour consumption, which fell to 137 pounds in 2002 after peaking in 1997 at 147 pounds, is a recent development. Declining consumption may reflect, in part, the increasing numbers of health-conscious people following low-carbohydrate diets, along with technological advances that extended the shelf life of bread, with less flour required to meet the same level of demand (Vocke et al., 2005).

Table 9. Number and Size of Plants, Labor Productivity, and Concentration in Soybean Processing, Flour Milling, and Fluid Milk Processing, 1972-2002

	1972	1977	1982	1987	1992	1997	2002
Soybeans							
Number of plants	94	121	114	106	99	93	117
Output per plant (million bushels)	7.6	7.8	9.8	11.1	12.9	17.2	13.8
Labor productivity (1,000 bushels)[1]	78	101	126	168	172	235	221
CR4[2]	52	50	61	71	71	79.6	79.9
Flour							
Number of plants	457	407	360	358	371	382	340
Output per plant (1,000 cwt)	548.0	677.6	808.1	954.1	999.5	1,055.2	1,160.9
Labor productivity (1,000 cwt)[1]	15.6	17.7	19.3	25.7	28.3	31.6	33.9
CR4[2]	32	33	40	44	56	48.4	53.6
Fluid milk							
Number of plants	2,507	1,924	1,190	946	779	608	524
Output per plant (million pounds)	Na	27.9	43.5	57.0	70.6	90.1	103.5
Labor productivity (1,000 pounds)[1]	Na	573	664	745	868	950	979
CR4[2]	17.0	17.0	16.0	21.0	22.0	21.3	42.6

[1] Annual output per employee.

[2] Four-firm concentration ratio.

Sources: Ollinger et al. (2005) and ERS calculations based on U.S. Census data.

Table 10. Change in Concentration, Sales (Volume) Growth, and Number of Plants and Firms, Selected Industries,1997-2002[1]

Type of Industry	CR4			Entire Industry		
	1997	2002	Amount of change	Sales (volume) growth[2]	Change in number of plants	Change in number of firms
	Percent			*Percent*		
11-15 percent increase in CR4:						
Cookie and cracker	60	67	11	0	-7	-9
Flour milling	48	54	11	-2	-11	-16
Dry pasta	57	65	13	-13	-27	-27
Poultry processing	41	46	14	16	14	21
Frozen fruit, juice, and vegetable	34	39	15	-9	-8	-15
16-30 percent increase in CR4:						
Commercial bakeries (L)	39	46	18	-6	-5	-6
Distilleries	60	71	18	-4	37	52
Meat processed from carcasses	20	24	19	5	3	3
Bottled water	52	63	21	59	66	84
>30 percent increase in CR4:						
Fats and oils, refining and blending	37	48	31	-6	-9	-16
Malt	69	91	32	-31	-21	-16
Dried and dehydrated food	30	42	40	62	17	18
Retail bakeries (L)	3	4	48	22	5	3

Table 10. (Continued)

	CR4			Entire Industry		
Type of Industry	**1997**	**2002**	**Amount of change**	**Sales (volume) growth[2]**	**Change in number of plants**	**Change in number of firms**
Ice cream and frozen dessert (L)	32	48	49	17	-9	-11
Fresh and frozen seafood	14	20	50	15	-10	-9
Ice (L)	24	43	76	24	-15	-17
Seafood canning	26	46	76	48	-16	-18
Fluid milk (L)	21	43	100	-2	-14	-22

CR4 = Four-firm concentration ratio. L = Local market industry.

[1] Industries listed are those with a change in the 4-firm concentration ratio (CR4) based on value of shipments, exceeding 10 percent from 1997 to 2002.

[2] Estimated from growth in the volume of shipments, deflated by producer price indexes, or taken directly from quantities obtained from other sources.

Sources: Compiled by ERS from U.S. Census Bureau data, Rodwan, 2003, *ERS Wheat Situation and Outlook*, *ERS Poultry Yearbook*, and NASS Agricultural Statistics for relevant years.

The fluid milk industry had the largest percentage increase in CR4 among all of the food and beverage industries. The industry's extraordinary increase in concentration has come amid dramatic structural changes in the dairy industry. In 2002, there were 33 percent fewer fluid milk processing plants than in 1992, processing 46 percent more milk per plant. Suiza Foods emerged as a prominent player in the 1990s by purchasing 20 fluid milk firms, while Dean Foods acquired 14 fluid milk firms in 1997 and 1998 (Ollinger et al., 2005). In 2001, a merger between Suiza Foods and Dean Foods, the Nation's number one and two dairy processors, resulted in a company accounting for 30 percent of the U.S. fluid milk market (USDA, 2004). Rapid consolidation in milk production was also taking place. From 1994 to 2004, the number of dairy farms decreased by 45 percent, but milk production per farm doubled (USDA/NASS).

Declining per capita consumption of fluid milk, which fell from 223 to 207 pounds over the 1994-2004 period, is one factor influencing consolidation. Other factors include scale economies and meeting the large-scale needs of an increasingly consolidated grocery retailing sector (USDA, 2004; Connor and Schiek, 1997).

Among the industries with increases in concentration exceeding 10 percent, bottled water was one of two with demand growth (i.e., sales volume growth) exceeding 50 percent. From 1997 to 2002, per capita consumption increased by 53 percent, from 14.1 to 21.5 gallons (Rodwan, 2003). U.S. consumers now drink more bottled water than any other beverage, with the exception of carbonated soft drinks (Beverage Marketing Corporation of New York, 2006). Since 1998, however, per capita consumption of carbonated soft drinks has fallen slightly, while bottled water consumption continues to set new records.

An increase in the concentration of bottled water production can be traced to a small number of companies following a cost-leader strategy. Diversification by firms into multiple product markets may affect the costs of entry when products are related through economies of scope in production, distribution, or marketing. Leveraged by their extensive bottling and distribution networks, Pepsi-Cola and Coca-Cola recently entered the bottled water industry and quickly ranked among the top four processors. The companies also relied on their own brand-building capabilities. Pepsi introduced Aqua-fina in the mid-1990s, and by 1999 the company had become the fourth- largest producer, with a 5.5-percent market share. In 1999, Coke introduced the Dasani brand, supported by the largest advertising budget in the industry (Beverage Marketing Corporation of New York, 2000). By 2002, market shares of Pepsi and Coke bottled water had reached 11 percent and 10 percent, respectively. The other two leading companies, European-based Nestle and Groupe Danone, also continued to grow market share through North American mergers and acquisitions and constructing plants.

Spotlight on the Pork and Beef Industries

Consolidation in beef and pork slaughter has been of special interest to policy officials, given the historically high and growing rates of concentration. The pork industry, in particular, has experienced rapid structural changes in recent years, epitomizing the industrialization of agriculture.

The U.S. Department of Agriculture provides information on market structure that is more recent and that enables more precise meat slaughter classifications than U.S. Census Bureau data. Fewer and larger hog slaughter plants continue to account for an increasing share of annual slaughter (table 11). The four largest hog slaughter firms have increased their share of slaughter by 20 percentage points over the past 10 years, reaching 64 percent in 2004. While this share remained stable from 1996 to 2002, it increased by 9 percentage points in 2003, following Smithfield Foods' acquisition of Farmland Foods (see the next section, "Recent Mergers, Acquisitions, and Divestitures in the Food Industry").

Increasing consolidation in hog slaughter is linked to dramatic structural changes in hog production. Since the 1980s, particularly since 1989, U.S. hog production has been shifting to highly specialized large-scale farms (Martinez et al., 1997). Hogs on farms with more than 1,000 head increased from 37 percent of the swine population in 1987 to 56 percent in 1994, and to 89 percent in 2004. From 1994 to 2004, the percentage of hogs produced on farms with 2,000 or more hogs increased from 38 to 79 percent (USDA/National Agricultural Statistics Service (NASS)).[18] Concentration in hog production, however, is considerably less than at the slaughter stage. In 2005, sows owned by the top four hog producers accounted for 23 percent of the U.S. breeding herd inventory (Freese, 2005; USDA/NASS).

Table 11. Decline in Hog and Cattle Slaughter Plants, Percent of Animals Slaughtered in Large Plants, and Slaughter Concentration, 1994 and 2004

	Number of slaughter plants			Percent slaughtered in large plants[1]		Share of slaughter by the 4 largest firms[2]	
	1994	**2004**	**Percent change**	**1994**	**2004**	**1994**	**2004**
Hogs	830	664	-20.0	62	78	44.5	64.1
Cattle	882	689	-21.9	46	52	68.7	69.2

[1] Large hog slaughter plants are those slaughtering at least 2 million head annually. Large cattle slaughter plants are those slaughtering at least 1 million head annually. The definition of large plants is not directly comparable for hogs and cattle because a 1-million-head cattle plant produces much more meat than a 2-million-head hog plant.

[2] Concentration in cattle slaughter is approximately 10 percentage points lower than for steer and heifer slaughter, but they exhibited similar trends over the past 10 years. The four-firm concentration ratio for steer and heifer slaughter, which accounts for about 80 percent of cattle slaughter, fell slightly, from 81.7 in 1994 to 79.6 in 2004.

Source: Compiled by ERS from USDA/NASS and USDA/GIPSA data.

Marketing contracts and packer-owned hogs are the preferred methods of vertical coordination between large packers and producers. In January 2006, 70 percent of all hogs were sold through marketing contracts, and 20 percent were owned and slaughtered by the same packer (Grimes and Plain, 2006). This suggests that only 10 percent of hogs are "negotiated purchase and negotiated sales," also referred to as spot market transactions, compared with 87 percent in 1993.[19]

In the cattle slaughter sector, relatively high concentration is coupled with continued consolidation into fewer and larger plants, albeit at slower pace than in the hog sector.[20] The decline in the number of cattle slaughter plants has matched the decline in hog plants. The share of hogs slaughtered by large plants (i.e., at least 2 million head annually) increased by 16 percentage points from 1994 to 2004, while the share of cattle slaughtered by large cattle slaughter plants (i.e., at least 1 million head annually) rose by 6 percentage points over the same period.[21] Unlike that of hog slaughter, the 4-firm concentration ratio for cattle slaughter has been fairly stable, hovering around 70 percent since the mid-1990s.

For the four largest steer and heifer packers, the percentage of slaughter that was contracted or packer-owned fell from 44.4 percent in 2002 to 38.5 percent in 2003. This was the first year since 1999, when 32.4 percent were contracted or packer-owned, that the share had fallen (USDA/GIPSA).

Recent Mergers, Acquisitions, and Divestitures in the Food System

This section highlights major mergers and acquisitions since 2002, the latest year available for Census industry concentration ratios. After an active period of food mergers and acquisitions from 1997 to 2001, merger activity in the food system declined beween 2002 and 2006 (figure 8). Manufacturers found performance results disappointing and decided to focus on their core assets and top-selling product lines. Recent major acquisitions, which include those in the meat and confectionery industries, have been as follows:

- In 2003, two notable meat-processor mergers occurred: Smithfield Foods, the world's largest pork packer and 10th-largest North American food processor, acquired the pork operations of Farmland Industries, referred to as Farmland Foods, the Nation's 8th-largest packer. Also, a group of investors led by U.S. Premium Beef acquired

Farmland National Beef, the fourth-largest beef processor, from Farmland Industries (American Institute of Food Distribution, Inc., *The Food Institute Report*, January 19, 2004).

- In 2004, the largest food-related deal was Wm. Wrigley Jr. Company's purchase of Kraft Food's confectionery brands, including Life Savers and Altoids, which was valued at $1.46 billion. The purchase was completed in 2005. Also in 2004, the 5th-largest pork processor, Hormel, acquired 11th-ranked Clougherty Packing.
- In 2005, the merger between beef packers American Foods Group Inc. and Rosen's Diversified, Inc., referred to as American Foods Group, LLC, solidified the sixth-ranked position of American Foods Group Inc.
- In 2006, mergers were announced between two leading chicken processors and between two leading pork processors. A $1.1 billion merger agreement between the second- and third-largest chicken processors, Pilgrim's Pride and Gold Kist, will create the world's largest chicken company, ahead of Tyson Foods. According to data contained in *Feedstuffs 2007 Reference Issue and Buying Guide*, the top four companies will account for 58 percent of chicken slaughter after the merger, up from 53 percent. Smithfield Foods, the largest pork processor, announced a proposed merger with Premium Standard Farms (PSF), the sixth-largest pork processor. Smithfield and PSF are also the largest and second-largest hog producers. If the merger is approved, the combined company will account for 20 percent of hog production and 31 percent of hog slaughter, and its own hogs will account for 54 percent of hogs slaughtered by the company (Kilman, 2006).

Major supermarket mergers and acquisitions have also slowed as companies integrate assets obtained in earlier merger activity. In 2003, most supermarket activity involved national chains who were attempting to increase their share in certain geographic markets through small acquisitions (American Institute of Food Distribution, Inc., January 19, 2004). In 2004, Albert-sons, the 3rd-largest grocery retailer, acquired JS USA Holdings, the 11th-ranked supermarket, which operates under the Shaw's and Star Markets banners.

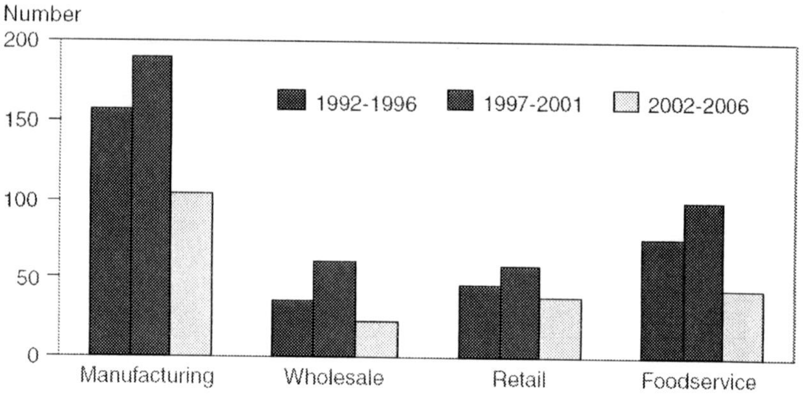

Source: Compiled by ERS from data from American Institute of Food Distribution, Inc.

Figure 8. Annual average mergers and acquisitions by type of food operation, 1992-2006

In 2003, the Nation's largest grocery wholesaler, Fleming, divested the bulk of its operations after its largest customer, Kmart, filed for bankruptcy protection in the face of stiff competition from Wal-Mart and Target.

Shortly after declaring bankruptcy in 2002, Kmart ended its supply contract with Fleming, which accounted for about 20 percent of Fleming's revenue. Fleming's financial problems also mounted because of competition from supercenters and conflicts with suppliers related to the company's restructuring efforts (Zwiebach, September 2005). In April 2003, Fleming filed for bankruptcy protection and subsequently closed distribution centers and divested the bulk of its food distribution and retail operations. Volume from Fleming's 32 distribution centers was replaced by over 100 suppliers in over 200 supply locations (Lempert, 2003). Following the divestitures, the share of general-line grocery sales by the top four wholesalers fell from 34 percent in 2001 to 27 percent in 2004 (Appendix A, Appendix table 3).

In 2006, the 11th-ranked grocery wholesaler, Associated Wholesalers, Inc. (AWI), acquired 9th-ranked White Rose Foods. The combined company is expected to benefit from complementary strengths in different product categories and to lessen AWI's exposure to Wal-Mart and other grocery retailers.

Given uncertainty that independent retailers served by general-line wholesalers can compete with larger chains, some of the largest grocery wholesalers are vertically integrating into retailing. For example, in 2005,

Roundy's, formerly the eighth-ranked grocery wholesaler, changed its name to Roundy's Supermarkets to better reflect the company's retail focus. The company also sold part of its distribution business to Nash Finch and Supervalu. In 2006, Supervalu, formerly the ninth-largest grocery retailer and second-largest grocery wholesaler, along with a consortium of investors that included CVS Corp., a leading drugstore chain, acquired Albertsons for $17.4 billion. The deal boosted Supervalu to the third-ranked position among grocery retailers, behind Wal-Mart and Kroger. This trend in vertical integration by wholesalers is expected to continue.[22]

Food Companies Expand Corporate Responsibility Reporting and Customized Marketing

Understanding, influencing, and meeting consumer demands are key functions of food marketing, with implications for the types of food products offered. Monitoring progress in these areas is important for appraising the outcomes of competitive strategies in the food marketing system. An intensely competitive environment increases the costs of competing, such as expenditures on new products and advertising. In this section, we describe some differentiation strategies used by food firms, including responsiveness to societal needs, introduction of new products, and advertising to influence consumption patterns.

The Corporate Social Responsibility Movement

One of the most striking global developments in business over the past 10 years has been the emergence of the "corporate social responsibility" (CSR) movement as part a firm's product differentiation strategy (Conley and Williams, 2005; Siegel and Vitaliano, 2006). The idea behind CSR is that the responsibility of a corporation extends beyond providing a financial return to its shareholders. Proponents of CSR argue that company objectives should broaden to include sustainable growth, equitable employment practices, and long-term social and environmental well-being. In addition, they believe that other groups should be included in corporate decisions, not only employees, but also residents affected by the decisions, governments, and organizations that are advocates for environmental and social causes. CSR shifts the emphasis from traditional government regulation of corporate conduct to the promotion of corporate disclosure of activities that address social and environmental issues. (See box, "The Five Stages of Corporate Responsibility," p. 31.)

One indication of the importance of the CSR movement in the United States is voluntary reporting by corporations of their social and environmental activities. Many of the largest companies have started producing social, environmental, or sustainability reports—a combination of social, environmental, and financial information. In 2005, 32 of the 100 top U.S. companies published stand-alone CSR reports (KPMG Global Sustainability Services, 2005). Many of the companies producing social and environmental reports have changed the way that they interact with nongovernmental organizations (NGOs), such as Greenpeace. Instead of relating as adversaries, they are partnering with NGOs to identify issues, conduct discussions with stakeholders around the world, produce and audit reports, and address specific problems.

The growing importance of CSR is also evident in the rise of M.B.A. programs that provide training in CSR. For example, more than half of the 100 M.B.A. programs surveyed worldwide by Aspen Institute and World Resources Institute require courses in corporate responsibility, compared with a third of M.B.A. programs in 2001 (Alsop, 2005). Major companies, such as McDonald's, have begun forging close relationships with business schools offering these courses. In 2005, McDonald's started a research fellowship at the University of California at Berkeley for six students to study social, environmental, and financial aspects of the beef supply chain.

Web sites and annual reports of the top five companies in main sectors of the U.S. food system, food manufacturers, foodservice companies, and grocery retailers, reveal substantial variation in the nature and extent of CSR reporting by food companies (table 12). At the time of this writing, all 15 of these companies did some CSR reporting. CSR activities were discussed under headings such as "Community," "Environment," "Nutrition, Health, and Wellness," and "Diversity." Eight companies provided both a Web site link devoted to CSR reporting and a dedicated CSR report, while two companies had only a devoted link. Reports were titled as either "corporate responsibility reports" or "sustainability reports," and ranged from 19 pages (ConAgra) to 88 pages (McDonald's). CSR reporting included discussions of company principles, goals, and accomplishments.

THE FIVE STAGES OF CORPORATE RESPONSIBILITY

What factors motivate food companies to engage in socially beneficial activities that extend beyond government mandates? In a 2005 *Harvard Business Review* excerpt, Zadek (2005) describes five stages as companies move toward greater social responsibility.

1. *Defensive stage*. Companies are faced with unexpected criticism from sources such as media and social activists. Typical response is to deny the allegations or a relationship between the company's practices and negative outcomes.
2. *Compliance stage*. Corporate policy is formed and observed, and is usually made visible to critics. Compliance is viewed as a cost of protecting the company's reputation and avoiding litigation. For example, the current public policy debate on obesity highlights food companies' aim for compliance, while the public expects far greater commitment.
3. *Managerial stage*. Companies realize that the problem cannot be deterred by simple compliance or public relations strategies.
4. *Strategic stage*. Companies learn how establishing strategies to address responsible business practices can give them a competitive advantage. One example is food companies that seek greater awareness of how their products affect consumer health.
5. *Civil stage*. Leading companies promote industry guidelines and actions to address societal concerns. For example, companies may engage in educational initiatives to promote healthy lifestyle choices, or refrain from sales promotion strategies that could adversely affect social welfare. In a 2005 press release, Kraft Foods announced an initiative to advertise more nutritionally healthy products in media primarily viewed by children ages 6-11.

Source: Zadek, 2005.

The extent of CSR reporting by leading companies, and the topic areas covered, varied by the company's role in the marketing chain. Ahold was the only grocery retailer with a link or report dedicated to CSR activities, whereas all five manufacturers and four foodservice companies had either a link or report. Furthermore, all five manufacturers had both a link and dedicated report, while only two foodservice companies did. More foodservice companies reported on animal welfare issues, while contributions to nutrition and healthy living were more commonly reported by both foodservice and manufacturing companies than by grocery retailers.

Table 12. Reporting of Corporate Social Responsibility (CSR) Activities by Leading U.S. Food Companies

Company	Web site link devoted to CSR reporting	Dedicated CSR report
Foodservice:[1]		
McDonald's	X	X
Yum! Brands	X	
Wendy's	X	
Burger King		
Starbucks	X	X
Retail:		
Wal-Mart		
Kroger		
Albertsons		
Safeway		
Ahold	X	X
Manufacturing:		
Tyson Foods	X	X
Kraft	X	X
Pepsico[2]	X	X
Nestle	X	X
ConAgra Foods	X	X

[1] Excludes Subway, which is a privately held company, owned by Doctor's Associates.
[2] A CSR report is published as part of the company's annual report.
Source: Compiled by ERS from company Web sites and annual reports.

The Global Reporting Initiative (GRI), which began in 1997, is an independent institution with the goal of developing guidelines for CSR reporting. Several of the leading food companies use the GRI index, which provides standardized guidelines for reporting progress on corporate economic, environmental, and social performance. Pepsico, Tyson Foods, and Ahold include information on the GRI index in their CSR reports, while Albertsons presents the index as part of its "Company Profile." According to

McDonald's "Worldwide Corporate Responsibility Report 2004," the company is committed to following GRI guidelines, but further work is needed on data collection systems and on improving the guidelines to better meet company needs. In its 2005 CSR report, Starbucks acknowledges not reporting in full accordance with GRI guidelines, but uses the principles and indicators as a basis for reporting. GRI indicators of a company's social responsibility commitment include its economic, environmental, labor, human rights, social (altruistic), and product responsibility practices.

New Product Introductions

New products bringing benefits to consumers, such as improved flavor or more convenient packaging, may indicate technological improvements and increased productivity. Food companies introduce new products to boost sales and maintain or expand market share. Product introductions may prevent competitors' new products from taking away sales or may stimulate demand in poorly performing categories. Retailers are more likely to allocate shelf space to manufacturers providing new products on a regular basis.

U.S. food and beverage product introductions tracked by Marketing Intelligence Service have been trending upward since the early 1990s, exceeding those of nonfood grocery items (figure 9).[23] A record 18,722 new food and beverage products were introduced in 2005 (table 13). Categories with the largest share of new products included candy, gum, snacks, beverages, condiments, and dairy. The increase in the share of new beverage products over the past 10 years is especially notable, along with the decline in the share of new condiments.

A growing share of new product introductions may suggest greater innovation for a category. However, new products may be variations of existing ones (e.g., new variety or package). Typically, over 90 percent of new food and beverage introductions are classified by Marketing Intelligence Service as "not innovative" (table 13). This suggests that food firms use new product introductions as a differentiation strategy, to offer a fresh image rather than truly novel benefits. Further, failure rates for new products are exceptionally high, exceeding 90 percent for some categories, which suggests that firms have difficulty in developing products that appeal to enough people to warrant continued distribution (Connor and Schiek, 1997). Nonetheless, some new products involve technological progress that delivers truly new benefits to consumers.

Advertising

Food advertising can influence consumption behavior by encouraging customers to try new products, maintaining brand loyalty, and capturing new customers (Connor and Schiek, 1997). Advertising expenditures as a percent of sales, or advertising intensity, decrease as one moves further down the supply chain (figure 10). Advertising expenditures by manufacturers have accounted for a growing share of sales since 1997, which is not the case for other participants in the food marketing system.

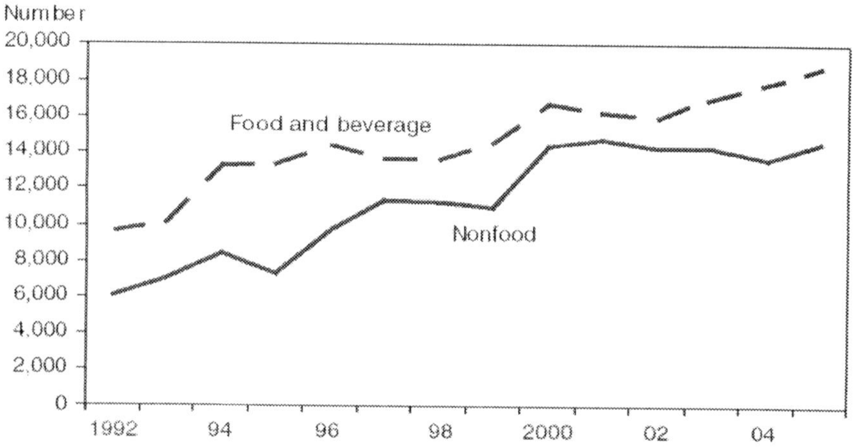

Note: Nonfood items include health and beauty aids, household products, pet products, and miscellaneous items (e.g., tobacco, car care, lighters).
Source: Marketing Intelligence Service, Ltd., Productscan Online, 2006.

Figure 9. New product introductions of consumer packaged goods, 1992-2005

The size of advertising expenditures by food companies indicates the degree of effort to influence consumer purchases. In 2004, food and beverage manufacturers ranked among the top 10 ad-spending industries, along with real estate agents, TV broadcast stations, and motion picture/video tape production (AdAge.com, 2005). In 2001, advertising expenditures by food and tobacco manufacturers accounted for 38 percent of all food and nonfood manufacturer advertising, up from 36 percent in 1994 (Internal Revenue Service, 2004, 1997). In 2004, several leading food manufacturers ranked among the top 50 companies in ad spending, including Altria Group, the parent company of Kraft Foods and Philip Morris ($1.4 billion), Pepsico ($1.3 billion), Nestle ($1 billion), Anheuser-Busch ($0.8 billion), and Mars ($0.7

Table 13. New Food and Beverage Product Introductions, Selected Years, 1995-2005

	1995	1998	2001	2004	2005
			Total		
New products	13,338	13,654	16,333	18,063	18,722
Type of product			*Percent of total*		
Candy, gum, snacks	25	23.8	32.5	30.5	27.7
Beverages	19.2	21.6	22.8	21.2	25.1
Condiments	18.2	14.2	13.8	13.9	10.2
Dairy	8.2	7.9	6.7	6.1	7.2
Baking ingredients	4.8	3.2	3.0	3.3	6.0
Processed meat	3.2	4.3	5.1	4.4	5.0
Meals and entrees	5.2	5.1	3.5	4.9	4.7
Bakery foods	5.6	5.2	2.8	5.3	4.1
Fruit and vegetables	3.0	5.1	2.8	3.3	3.4
Pasta and rice	2.9	4.8	2.3	2.2	2.2
Soups	1.9	2.1	1.8	1.5	1.6
Cereals	1.1	1.4	1.0	1.2	1.4
Desserts	.7	.5	.7	.6	0.8
Meal replacements and special diet foods	.6	.6	.5	.8	0.4
Baby food	.3	.2	.7	.7	0.3
Classification:					
Innovative[1]	8.4	7.1	7.4	8.8	5.4
Formulation[2]	5.2	4.4	4.9	6.1	3.4
Merchandising[3]	0	.1	.3	.4	0.3
New market[4]	0	0	0	0	0.0
Packaging benefit	1.4	.9	1.4	1.5	0.9
Positioning[5]	1.8	1.7	.7	.6	0.5
Technology[6]	0	0	.1	.2	0.3
Not innovative	91.6	92.9	92.6	91.2	94.6

[1] Products classified as innovative are those deemed by Marketing Intelligence Service to be the first to offer breakthrough features and benefits. Examples include icing marketed in microwaveable pouches, and a hot sauce with an adjustable cap that can be used to vary the spiciness.

[2] Products containing new ingredients that offer benefits not previously provided.

[3] Products marketed in a new way, such as unique display or packaging options.

[4] New products that do not compete with any existing product categories.

[5] New products presented for new users or uses compared with existing products.

[6] New product with added consumer benefits resulting from the new technology.

Source: Marketing Intelligence Service, Productscan Online, 2006.

billion). Several retailers and foodservice companies also ranked among the top 50, including McDonald's ($1.4 billion), Wal-Mart ($0.8 billion), and Yum Brands ($0.8 billion), whose subsidiaries include KFC, Pizza Hut, and Taco Bell.

In 2004, manufacturers of food and beverages (excluding alcohol) spent by far the largest share of their media budgets on TV advertising (69 percent), followed by ads in magazines (25 percent) and on the radio (2 percent) (*Advertising Age*, 2005). However, the share spent on TV advertising has declined, while magazine ads have increased: in 1990, manufacturers spent 80 percent of their media advertising for brand foods on TV ads, with 11 percent going to magazine ads (Connor and Schiek, 1997). The decline may reflect new TV commercial-skipping technologies, such as TiVo, and fragmentation of consumers across a wide variety of advertising venues, including cable TV, Internet sites, and video games (Vranica, 2005). Alternative advertising venues may improve the ability to identify and reach target customers. In 2004, cable networks accounted for 19 percent of the $6.8 billion in measured media advertising expenditures by food and beverage manufacturers (*Advertising Age*, 2005).

The Internet accounts for only 1 percent of manufacturer advertising expenditures, but it is becoming increasingly important in targeting children (*Advertising Age*, 2005; Shields, 2005). Online advertising to children generally takes the form of games and contests built around the brands and their characters. On Wrigley's "Bubblegum.com" stand-alone gaming site, young viewers can play carnival toss with "Hubba Bubba" bubblegum characters. Other examples include Post's Postopia.com and Frito-Lay's INNW.com.

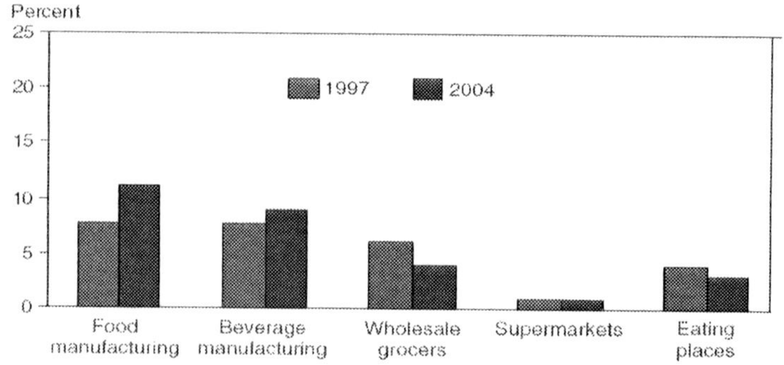

Source: AdAge.com.

Figure 10. Advertising expenditures by food companies as a share of total sales, 1997 and 2004

Some companies are turning to video games, in-store advertising, and product placement in entertainment programs to tout their brands (Nelson and Ellison, 2005; Turner, 2005; Vranica, 2005). McDonald's uses video games to target 18- to 34-year-old males, who are spending less time watching TV. Kroger announced plans to offer in-store television broadcasts of promotions and new products that will allow manufacturers to air their commercials. Each store will have programs specific to its location. In 2005, 14 percent of retailers, including Wal-Mart, were already using this technology, and 30 percent intended to add screens within 2 years. Company brands are also appearing in films. For example, Pepsico's Mountain Dew financed a documentary on snowboarding, and the brand could be seen occasionally in the movie.

Advertising messages that tout a product's attributes are also conveyed on packages and in supporting literature. Based on new product tags or claims tracked by Marketing Intelligence Service in 2005, health- and convenience-related attributes accounted for 9 of the top 10 subject categories for ads on packages (table 14).[24] Six of these categories, including those stressing "natural," "organic," "single serving," "quick," "fresh," and "low or no fat," have ranked among the top 10 in every year since 2001.

The low-carbohydrate craze is testimony to the responsiveness of the food industry to changes in consumer preferences, as well as to the difficulty of anticipating consumer demands. Given the relative ease of altering products for the "low-carb" market, the number of companies offering low-carb-labeled products grew rapidly to include mainstream players such as Nestle, Sara Lee, Frito-Lay, and the Nation's second-largest food company, Kraft. Supermarkets also responded, with shelf tags, special low-carb sections, features in weekly circulars, and other advertising. In 2004, new product introductions with low- or no-carb labels rose to 983, or 5.5 percent of all new food and beverage product introductions, compared with 209 (1.3 percent) in 2003 and 101 (0.9 percent) in 2001 (Marketing Intelligence Service, Ltd., 2006). More recently, the popularity of the Atkins and other low-carb diets has waned, punctuated by the July 2005 bankruptcy of Atkins Nutritionals. In 2005, 399 "low- or no-carb" products were introduced, down 60 percent from 2004.[25]

While the low-carb craze is apparently fading, a Federal law requiring manufacturers to list trans-fat on food labels by 2006 has sparked reformulation of products to reduce trans-fat. In 2005, 458 "no- or low-trans-fat" products were introduced, compared with 238 in 2004, 64 in 2003, and 5 in 2001 (Marketing Intelligence Service, Ltd., 2006). This reformulation is likely to continue as processors work with farmers and suppliers to increase production of low-linolenic soybean oil, and other healthier cooking oils. which can be used to reduce trans-fatty acids (just-food.com, 2005).

Table 14. Number of New Product Introductions in the Top 10 Product Claim Categories for 2001 to 2005[1]

Tag or claim[2]	2001	2002	2003	2004	2005
	Number				
Upscale	695	905	1589	1,568	2,106
Natural	1,063	1,245	1,380	1,364	1,612
Single serving	782	1,119	1,127	1,111	1,277
Fresh	479	578	556	599	692
Organic	378	443	559	533	670
Low or no fat	518	586	656	617	646
Quick	377	443	521	511	571
No preservatives	368	421	578	551	550
Low or no sugar	227	283	395	619	544
High-vitamin	381	422	483	509	532
Total new product claims	11,149	13,769	16,374	17,922	19,544
	Percent of total				
Upscale	6.2	6.6	9.7	8.7	10.8
Natural	9.5	9.0	8.4	7.6	8.2
Single serving	7.0	8.1	6.9	6.2	6.5
Fresh	4.3	4.2	3.4	3.3	3.5
Organic	3.4	3.2	3.4	3.0	3.4
Low or no fat	4.6	4.3	4.0	3.4	3.3
Quick	3.4	3.2	3.2	2.9	2.9
No preservatives	3.3	3.1	3.5	3.1	2.8
Low or no sugar	2.0	2.1	2.4	3.5	2.8
High-vitamin	3.4	3.1	2.9	2.8	2.7

[1] Does not include associated stock keeping units (SKUs) (e.g., variations in size and form). According to Marketing Intelligence Service, the SKU count report may produce erroneous results because a single new product introduction can have multiple SKUs and each of these SKUs may or may not have certain package tags.

[2] A new product may have multiple tags or claims.

Source: Marketing Intelligence Service, Productscan Online, 2006.

Through an agreement with Monsanto, a leading supplier of agricultural products, Kellogg will become one of the first food manufacturers to use low-linolenic soybean oil. To supply low-linolenic soybeans, farmers will need to change their production methods. Kellogg is encouraging farmers to produce the soybeans under contract with participating processors. The success of specialty soybean marketing will hinge on several factors, including market demand and volume, identity preservation, access to appropriate distribution channels, and mutual trust, cooperation, and information-sharing at each level of the supply chain (Darroch et al., 2002).

"Upscale" ranked as the leading new product claim category in 2003, 2004, and 2005, surpassing "natural" and "single serving." Examples include premium ice cream, uniquely processed coffee, gourmet jam and dessert topping, and Certified Angus Beef. In what sociologists refer to as a "post-modern consumer culture," the purchase of upscale food products allows consumers to express their individuality (Kinsey, 1994; Krasteva, 2003). As incomes rise and consumers continually seek new experiences and tastes, the market grows for new niche products that allow consumers to signify their independence and social position. Upscale products create an endless stream of niche-marketing opportunities for food companies, as new products must be continually introduced to replace those that become more mainstream. This reinforces the trend away from traditional mass marketing toward differentiated and customized products.

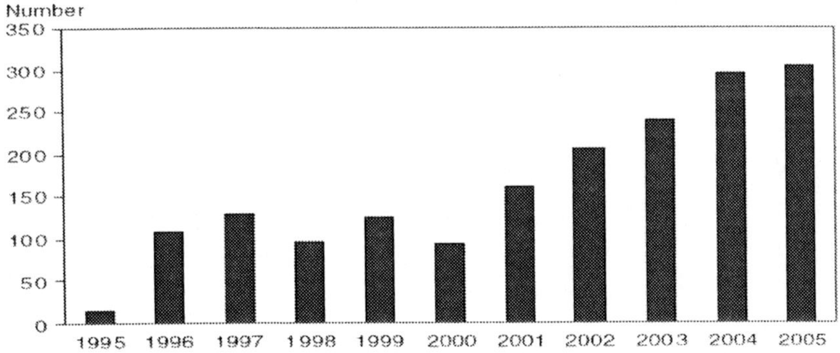

*In cobranding, a company that produced a familiar ingredient used in a brand-name product is paid by the maker of the product to have the logo of the ingredient displayed on the package.

Source: Marketing Intelligence Service, 2006.

Figure 11. Number of new cobranded* products, 1995-2005

Advertising through "cobranding" has become an increasingly popular strategy for differentiating food products. Food processors typically pay a fee or royalty to place the logo of a popular food ingredient, container type, or media character—especially one that appeals to children—on a brand-name package (Connor and Schiek, 1997). The logos represent a strong image that is well recognized by consumers. In 2005, 302 cobranded products were introduced, compared with 16 in 1995 (figure11). Recent examples include products containing *Splenda* brand sweetener and carrying Splenda's logo, Tops brand office cream containing *Nestle Butterfinger* candy pieces, Hillshire Farm deli products packaged in *Glad Ware* containers, and *SpongeBob Squarepants* characters on packages of Keebler Animal Crackers.

Appendix A. Top 10 Companies in Each Stage of the Food Marketing System and Changes in Concentration in all Food and Beverage Manufacturing Industries and Grocery Wholesaling

Appendix Table 1. Top 10 Food Marketing Companies by Stage of Marketing[1]

Manufacturing[2]	Grocery wholesalers[3]	Foodservice wholesalers	Grocery retailers	Foodservice
Tyson Foods, Inc. ($23.9)	C and S Wholesale Grocers ($15.2)	Sysco Co. ($31.4)	Wal-Mart Stores ($98.7)	McDonald's Corp. ($26.9)
Kraft Foods, Inc.[4] ($23.3)	Supervalu, Inc. ($10.00)	U.S. Foodservice ($18.5)	Kroger ($58.5)	Yum! Brands[5] ($17.4)
Pepsico, Inc. ($21.2)	Wakefern Food Corporation ($9.0)	Performance Food Group ($5.7)	Albertsons, Inc. ($36.3)	Wendy's International, Inc. ($8.0)
Nestle ($19.9)	Associated	Gordon Food	Safeway, Inc.	Burger King
	Wholesale Grocers of Kansas City ($4.5)	Service ($3.7)	($32.7)	($7.9)

Appendix Table 1. (Continued)

Manufacturing[2]	Grocery wholesalers[3]	Foodservice wholesalers	Grocery retailers	Foodservice
Anheuser-Busch Cos., Inc. ($11.5)	Nash Finch Company ($3.9)	Food Services of America Inc. ($2.5)	Ahold USA, Inc.[6] ($23.8)	Doctor's Associates, Inc.[7] ($7.2)
Dean Foods ($10.5)	Unified Western Grocers ($2.9)	Reinhart Foodservice, Inc. ($2.1)	Publix Super Markets, Inc. ($18.5)	Starbucks Corp. ($5.8)
General Mills, Inc. ($9.8)	United Natural Foods ($2.3)	Maines Paper and Foodservice, Inc. ($2.0)	Delhaize America, Inc.[8] ($16.5)	Darden Restaurants, Inc.[9] ($4.8)
Smithfield Foods, Inc. ($9.6)	Purity Wholesale Grocers ($1.2)	Ben E. Keith Foods ($1.3)	H.E. Butt Grocery Co. ($10.4)	Allied Domecq[10]($4.5)
ConAgra Foods, Inc. ($8.2)	White Rose Foods ($1.2)	Shamrock Foods Co. ($1.3)	Supervalu, Inc. ($8.6)	Applebee's Internationa, Inc. ($4.2)
Swift & Co. ($7.8)	Spartan ($1.2)	Cheney Brothers, Inc. ($0.6)	Winn-Dixie Stores, Inc.($7.1)	Brinker International[11] ($4.2)

[1] 2005 sales in parentheses (billion dollars).
[2] Includes North American food sales.
[3] Excludes sales at company-owned retail stores.
[4] Subsidiary of Altria Group.
[5] Chains include KFC, Pizza Hut, Taco Bell, and Long John Silver's.
[6] Stores include Giant Food, Stop and Shop, and Tops Friendly Markets.
[7] Owner of Subway.
[8] Stores include Food Lion, Hannaford Bros., and Kash 'n Karry.
[9] Chains include The Olive Garden and Red Lobster.
[10] Chains include Dunkin' Donuts and Baskin-Robbins.
[11] Chains include Chili's Grill & Bar, Romano's Macaroni Grill, and On The Border.

Sources: Compiled by ERS from *Food Processing*, August 2006; *Supermarket News*; American Institute of Food Distribution, Inc., *The Food Institute Report*, April 3, 2006; *Progressive Grocer*, May 2006; Technomic Information Services, Inc., 2006; and company annual reports filed with the SEC.

Appendix Table 2. Changes in Number of Plants, Number of Firms, and Concentration in Food and Beverage Manufacturing, 1997 to 2002

	Number of plants	Number of firms	CR4[4]
	Percent change		
Average change over all industries	4.5	6.3	11.4
Increasing concentration			
Fluid milk	-14	-22	100
Seafood canning	-16	-18	76
Ice	-15	-17	76
Fresh & frozen seafood	-10	-9	50
Ice cream and frozen desert	-9	-11	49
Retail bakeries	5	3	48
Dried & dehydrated food[2]	17	18	40
Malt	-21	-16	32
Fats & oils refining & blending	-9	-16	31
Bottled water	66	84	21
Meat processed from carcasses	3	3	19
Commercial bakeries	-5	-6	18
Distilleries	37	52	18
Frozen fruit, juice, & vegetable	-8	-15	15
Poultry processing	14	21	14
Dry pasta	-27	-27	13
Cookie & cracker	-7	-9	11
Flour milling	-11	-16	11
Dog & cat food	29	36	10
Nonchocolate confectionery	-17	-18	10
Creamery butter	3	3	10
Other animal food[1]	4	8	10
Soft drink	-16	-24	10
Wineries	74	77	8
Flour mixes & dough mfg. from purchased flour	-12	-13	8
Other snack food[3]	-19	-22	6
Specialty canning	-7	-6	6
All other miscellaneous food	96	102	4
Animal (except poultry) slaughtering	34	36	3
Rice milling	13	11	3
Frozen specialty food	1	-1	2
Dry, condensed, & evaporated dairy product	0	-5	1

Appendix Table 2. (Continued)

	Number of plants	Number of firms	CR4[4]
Breweries	-29	-30	1
No change in concentration			
Cheese	-4	-8	0
Soybean processing	27	42	0
Reduction in concentration			
Tortillas	27	25	-2
Coffee & tea	14	20	-2
Perishable prepared food	38	35	-3
Wet corn milling	5	10	-4
Mayonnaise, dressing, & other prepared sauce	-8	-11	-4
Fruit & vegetable canning	-5	-4	-5
Breakfast cereal	-7	-4	-5
Beet sugar	-3	38	-6
Confectionery mfg'd. from purchased chocolate	26	27	-7
Sugarcane mills	-16	-15	-9
Rendering & meat byproduct processing	-3	-11	-9
Roasted nuts & peanut butter	13	11	-11
Chocolate & confectionery mfg'd. from cacao beans	-8	-10	-13
Other oilseed processing	-22	9	-14
Frozen cakes, pies, & other pastries	10	1	-21
Spice and extract	12	14	-34
Not disclosed			
Flavoring syrup & concentrate	-7	-3	ND
Cane sugar refining	11	17	ND

[1] Animal food, except dog and cat.

[2] Includes mixing of purchased dried and/or dehydrated ingredients for such products as soup mixes and bouillon.

[3] Excludes roasted nuts and peanut butter.

[4] Concentration as measured by the percent of shipment value accounted for by the top 4 companies.

Source: Compiled by ERS from U.S. Census Bureau data.

Appendix Table 3. Top 10 Wholesalers Ranked by Wholesale Sales, 2001 and 2004

Company	2001 Sales derived from			Company	2004 Sales derived from		
	Wholesale	Retail	Total		Wholesale	Retail	Total
	Billion dollars				*Billion dollars*		
Fleming Companies, Inc.	13.3	2.3	15.6	C and S Wholesale Grocers	13.2	0.0	13.2
Supervalu, Inc.	10.8	9.5	20.3	Supervalu, Inc.	9.0	10.5	19.5
C and S Wholesale Grocers[1]	8.5	0.0	8.5	Wakefern Food Corporatio n	7.0	0.1	7.1
Wakefern Food Corporation	5.8	0.1	5.9	Associated Wholesale Grocers of Kansas City	4.3	0.5	4.8
Nash Finch Company	3.1	1.0	4.11	Nash Finch Company	3.1	0.8	3.9
Associated Wholesale Grocers (Kansas City)	3.0	0.2	3.1	Unified Western Grocers	3.0	0.0	3.0
Unified Western	2.9	0.2	3	Giant Eagle Grocers	1.9	3.3	3.0
Spartan Stores, Inc.	2.1	1.4	3.5	Roundy's, Inc.	1.8	3.0	4.8
Roundy's, Inc.	2.0	1.4	3.4	Purity Wholesale Grocers	1.4	0.0	1.4
Giant Eagle	1.6	2.9	4.5	Associated Wholesaler	1.3	0.1	1.4
Total grocery general line merchant Wholesalesales	112.3				122.7		
Top 4 percent of total	34.2				27.3		

[1] Excludes sales at its Grand Union stores.

[2] Excludes manufacturing sales branches and offices. Because general-line grocery wholesale sales, excluding specialty and limited-line whole-salers, are not available in non-Census years, they were assumed to account for 30 percent of merchant sales. This percentage is based on the 2002 census year.

Sources: *Supermarket News,* U.S. Census Bureau, and annual reports filed with the SEC.

APPENDIX B. MAJOR CHANGES IN FOOD INDUSTRY CENSUS CLASSIFICATIONS IN 1997 AND 2002

In 1997, the U.S. Census Bureau revised the coding system for industry classifications from the Standard Industrial Classification (SIC) code to the North American Industrial Classification System (NAICS). Major changes affecting the food industry include:

- Retail bakeries that make and sell products at the same location, and candy stores that make candy on the premises, which had been classified as retail establishments, are now classified as manufacturers.
- Cookie and donut shops, formerly classified as retail establishments, are now classified as "snack and nonalcoholic beverage bars" under the NAICS code for "foodservices and drinking places."
- Gas stations with convenience stores, which had been classified as food-stores, are now part of the gas station industry.
- Beer, wine, and liquor stores were moved from the "miscellaneous store" classification to a new "beer, wine, and liquor stores" classification, replacing the old SIC code for liquor stores.
- Beverage manufacturing was moved from the "food and kindred products" SIC code to a new "beverage and tobacco product" NAICS code.

In 2002, the NAICS codes were further refined, with significant changes in wholesaling.[26] Under the 1997 classifications, agents and brokers, who do not take title to the goods being sold but rather receive a commission or fee for their service, were classified as wholesalers in the "grocery and related products" industry. Beginning in 2002, agents and brokers were moved to the "wholesale electronic markets and agents and brokers" industry segment. In addition to agents and brokers, the wholesale sector is comprised of merchant wholesalers that sell products on their own account, including:

- Manufacturers' sales branches and offices, which are maintained by manufacturers for marketing their products.
- Merchant wholesalers, excluding manufacturers' sales branches and offices.

Because 1997 was the first year of the new classification system, 1997 and 2002 data were used to analyze changes in the number and size of companies and establishments.

References

Adage.com. (2004). Advertising to Sales Ratios for 200 Largest Ad Spending Industries," accessed March 16, 2005 at: www.adage.com/ page.cms? pageId=1064/.

Adamy, J. (2005). "New Food, New Look," *The Wall Street Journal*, November 21.

Advertising Age. (2005). "Domestic Advertising Spending by Category," June 27.

Alexander, D. (2004) "Grocers, Drugstores Mixing it up in Food Fight," *Chicago Tribune*, March 14.

Alsop, R. (2005). "Recruiters Seek M.B.A.s Trained in Responsibility," *The Wall Street Journal*, December 13.

American Institute of Food Distribution, Inc. *The Food Institute Report*. Selected issues. Elmwood Park, NJ.

Besanko, David, David Dranove & Mark Shanley (1996). *Economics of Strategy*. New York, NY: John Wiley and Sons.

Beverage Marketing Corporation of New York. "*Bottled Water Continues Tradition of Strong Growth in (2005)*," Press release, April 2006.

Beverage Marketing Corporation of New York (2000). "*U.S. Bottled Water Sales Spurt Forward, Recording Fastest Growth in Ten Years*," Press release, August 25.

Bjornson, B. & Sykuta, M. E. (2002). "Growth by Acquisition and the Performance of Large Food Retailers," *Agribusiness*, *Vol. 18*, pp. 263-81.

Brady, R. & Cavanaugh, K. (2005). "Grocery and Consumables Retail Landscape Continues to Evolve," *Competitive Edge*, June 2005 (revised July 14,).

Caffarini, A. & Cavanaugh. "The Future of Food Retailing," *Competitive Edge*, July 2006.

Collins, J. (2006). "RFID's Impact at Wal-Mart Greater Than Expected," *RFID Journal*, May 4.

Conley, J. M. & Williams, C. A. (2005) "Engage, Embed, Embellish: Theory Versus Practice in the Corporate Social Responsibility Movement." University of North Carolina (UNC) Legal Studies Research Paper No. 05-16, March 23. Available at SSRN: http://ssrn.com/abstract= 691521 or DOI: 10.2139/ssrn.691521.

Connor, John M. & William, A. Schiek. (1997). *Food Processing: An*

Industrial Powerhouse in Transition. New York, NY: John Wiley and Sons.

ConsumerReports.org. August (2005). "Store Brands vs. Name Brands: How to Choose," accessed September 18, 2006 at: http://www.consumer reports.org/ cro/personal-finance/store-brands-vs-name-brands-how-to-choose-805/index.htm .

Covert, J. (2005). "Wal-Mart Says Radio Tags Keep Shelves Better-Stocked," *The Wall Street Journal*, October 26.

Darroch, M. A., Akridge, J. T. & Boehlje, M. D. (2002). "Capturing Value in the Supply Chain: The Case of High Oleic Acid Soybeans," *International Food and Agribusiness Management Review*, *Vol. 5*, pp. 87-103.

Enis, M. (2005). "Quick-Service Restaurants Make the Cut," *Supermarket News*, September 26.

Feedstuffs. (2007) Reference Issue and Buyers Guide, Rural Press Limited, September 13, 2006.

Food Marketing Institute. (2005). "New FMI Report Shows Retailer Success With Sunrise 2005 and Data Synchronization; Barriers Remain for Electronic Product Code," Press release, September 15.

Forbes.com. (2004). "Emerging Trends: Private Labels Outperform Familiar Names," accessed November 18, 2005 at: http://www. forbes.com/ 2004/07/07/0707findspv_labels_print.html/ July 7.

Freese, B. (2005). "Show Me the Sows," *Successful Farming*, October.

Garry, M. (2005). "A Brave New World of Data Sync," *Supermarket News*, January 17.

Garry, M. "Bracing for Competition," (2005). *Supermarket News*, January 31.

Garry, M. "Data Collaboration," (2005). *Supermarket News*, June 20.

Garry, M. "Pursuing the Dream," (2006). *Supermarket News*, June 5.

Garry, M. (2005). "The New High-Tech Store," *Supermarket News*, November 28.

Garry, M. (2005). "Wakefern Chooses Transora as Data Pool," *Supermarket News*, March 14.

Global Insight Inc. (2005). "The Economic Impact of Wal-Mart." accessed December 13, 2006 at: http://www.globalinsight. com/MultiClientStudy/MultiClientStudyDe-tail2438.htm / November 2.

Greenhouse, S. (2004). "Wal-Mart, a Nation Unto Itself," *The New York Times*, April 17.

Griffith, T. (2002). "The U.S. Retail Food Industry: 2002 Store Format Update," *Competitive Edge*, June.

Grimes, G. & Plain, R. (2006). "*U.S. Hog Marketing Contract Study*," Department of Agricultural Economics, University of Missouri, Columbia, Working Paper No. AEWP 2006-01, January.

Hamstra, M. (2005). "Alternative Access: Club Stores," *Supermarket News*, October 24.

Harris, J. M., Kaufman, P. R. Martinez, S. W. & Price. C. (2002). *The U.S. Food Marketing System, 2002: Competition, Coordination, and Technological Innovations Into the 21st Century*, AER-811, U.S. Department of Agriculture, Economic Research Service, June.

Hopkins, J. (2003). S. Economy Follows the Wal-Mart Way," *USA Today*, February 3.

Internal Revenue Service. (2004). *Statistics of Income - 2001, Corporation Income Tax Returns*, Publication 16, September.

Internal Revenue Service. (1997). *Statistics of Income - 1994, Corporation Income Tax Returns*, Publication 16, September.

Irwin, E. G. & Clark. J. (2006). "The Local Costs and Benefits of Wal-Mart." Department of Agricultural, Environmental, and Development Economics, The Ohio State University. Accessed October 11, 2006 at: http://aede.osu.edu/programs/ComRegEcon/retail.htm/.just-food.com February 23.

just-food.com. (2005). "*Kellogg Moves to Low Linolenic Oil to Cut Trans Fat*," December 9.

Kaufman, P. R. (2000). "Consolidation in Food Retailing: Prospects for Consumers and Grocery Suppliers," *Agricultural Outlook*, August.

Kilman, S. (2006). "Smithfield To Buy Hog Farmer Premium Standard," *The Wall Street Journal*, September 19.

Kinsey, Jean. (1994). "Changes in Food Consumption from Mass Market to Niche Markets," *Food and Agricultural Markets: The Quiet Revolution*. National Planning Assoc. Rpt. No. 270. Eds.: Lyle P. Schertz and Lynn M. Daft. Washington, DC, pp. 19-43.

Kinsey, J. D. (2001). "The New Food Economy: Consumers, Farms, Pharms, and Science," *American Journal of Agricultural Economics, Vol. 83*, pp. 1113-30.

Kinsey, J. (2000). "Retail Food Distribution: The Race to Efficiency." Presentation at the International Institute of Fisheries Economics and Trade conference, Corvallis, OR July.

KPMG Global Sustainability Services. (2005). *KPMG International Survey of Corporate Responsibility Reporting 2005*, Report 15-6-2005, June.

Knowledge@Wharton. (2005). "*Spreading Yourself Too Thin: The Atkins Diet and Other Fads*," accessed September 13, 2006 at: http://knowledge.wharton.upenn.edu/article.cfm?articleid=1295&CFID=9 78 232&CFTOKEN=20486806/ October 19.

Krasteva, M. (2003). "*Creating a Postmodern Individual Product Offer*." Masters thesis, School of Economics and Commercial Law, Göteborg University, Göteborg, Sweden.

Lempert, P. (ed.). (2003). *Facts, Figures, and the Future*, accessed June 22, 2005 at: www.factsfiguresfuture.com/archive/november_2003.htm/ November 12.

Lempert, P. (2005). "The Lempert Report: Organic Generation," *Progressive Grocer*, October 1.

Leibtag, E. (2006). *The Impact of Big-Box Stores on Retail Food Prices and the Consumer Price Index*, ERR-33, U.S. Department of Agriculture, Economic Research Service, December.

MacDonald, J. M. & Ollinger, M. E. (2005). "Technology, Labor Wars, and Producer Dynamics: Explaining Consolidation in Beefpacking," *American Journal of Agricultural Economics*, *Vol. 87*, pp. 1020-33.

MacDonald, J. M., Ollinger, M. E. Nelson, K. E. & Handy C. R. (1999). *Consolidation in U.S. Meatpacking*, AER-785, U.S. Department of Agriculture, Economic Research Service, March.

Marketing Intelligence Service, Ltd. Productscan Online, 2006. http://www.productscan.com/.

Martinez, S. W. (1999). *Vertical Coordination in the Pork and Broiler Industries: Implications for Pork and Chicken Products*, AER-777, U.S. Department of Agriculture, Economic Research Service, April.

Martinez, S., Smith, K. & Zering, K. (1997). *Vertical Coordination and Consumer Welfare: The Case of the Pork Industry*, AER-753. U.S. Department of Agriculture, Economic Research Service, August.

Martinez, S. & Stewart. H. "From Supply Push to Demand Pull: Agribusiness Strategies for Today's Consumers," *Amber Waves*, *Vol. 1*, Issue 5, November 2003.

Martinez, S. W. & Zering, K. (2004). *Pork Quality and the Role of Market Organization*, AER-835, U.S. Department of Agriculture, Economic Research Service, October.

McBride, W. D. & Key (2003). *Economic and Structural Relationships in U.S. Hog Production*, AER-818. U.S. Department of Agriculture, Economic Research Service, February.

McKinsey Global Institute. (2001). *U.S. Productivity Growth 1995-2000*, Washington, DC, October.

Mohtadi, Hamid & Jean D. Kinsey. (2005). "Information Exchange and Strategic Behavior in Supply Chains: Application to the Food Sector," *American Journal of Agricultural Economics*, *Vol. 87*, pp. 582-99.

Muirhead, S. (2005). "McDonald's Rolls Out New Labels," *Feedstuffs*, October 31.

Nelson, E. & Ellison, S. (2005). "In a Shift, Marketers Beef Up Ad Spending Inside Stores," *The Wall Street Journal*, September 21.

Ollinger, M., Nguyen, S. V., Blayney, D., Chambers, B. & Nelson, K. (2006). *Food Industry Mergers and Acquisitions Lead to Higher Labor Productivity*, ERR-4227. U.S. Department of Agriculture, Economic Research Service, October.

Ollinger, M., Nguyen, S. V., Blayney, D., Chambers, B. & Nelson, K. (2005). *Structural Change in the Meat, Poultry, Dairy, and Grain Processing*

Industries, ERR-423. U.S. Department of Agriculture, Economic Research Service, March.

Parks, L. (2005). "More Retailers, Vendors Start Embracing RFID," *Supermarket News*, Sept. 26.

Porter, Michael, E. (1990). *The Competitive Advantage of Nations*. New York, NY: The Free Press.

Private Label Manufacturers Association. (2006). "Market Profile," accessed September 19, 2006 at: http://plma.com/store Brands/ sbt06.html.

Progressive Grocer. (2006). "Taking the Lead," May 1.

Rodwan, J. G. (2003). "The 2002 Stats: Another Year of Big Gains," *Bottled Water Reporter*, April/May.

Rogers, R. T. (2000). "The U.S. Food Marketing System." In *Wiley Encyclopedia of Food Science and Technology*, ed. F.J. Francis, pp. 2701-24. New York, NY: John Wiley and Sons.

Rogers, T. K. (2005). "Turning Supermarkets Into Restaurants, Too," *The New York Times*, August 28.

Seifert, Dirk. (2002). *Collaborative Planning, Forecasting and Replenishment*. Bonn, Germany: Galileo Press.

Sharkey, T. & Stiegert. K. (2006). "Impacts of Nontraditional Food Retailing Supercenters on Food Price Changes," Food System Research Group, Department of Agricultural and Applied Economics, University of Wisconsin-Madison, FSRG Monograph Series No. 20. February.

Shields, M. (2005). "Web Marketing to Kids Is Rising," *Mediaweek*, accessed September 19, 2006 at: http://www.aef.com/ industry/news/ data/2005/3137 July 25.

Showalter, K. (2006). "Grocers See Gold in Gas, Add Discounted Pumps," *Business First of Columbus*, June 12.

Siegel, D. S. & Vitaliano, D. F. (2006). "An Empirical Analysis of the Strategic Use of Corporate Social Responsibility." Rensselaer Working Papers in Economics 0602, Rensselaer Polytechnic Institute, Department of Economics. January.

Sparks, L. & Wagner. B. A. (2003). "Retail Exchanges: A Research Agenda," *Supply Chain Management: An International Journal*, *Vol. 8*, pp. 17-25.

Stewart, H. & Martinez. S. (2002). "Innovation by Food Companies Key to Growth and Profitability," *FoodReview*, *Vol. 25*, Issue 1, Spring.

Sullivan, L. (2005). "Wal-Mart RFID Trial Shows 16% Reduction in Product Stock-Outs," *RFIDinsights*, October 14.

Tarnowski, J. (2005). "Natural Defenses," *Progressive Grocer*, March 1.

Technomic, Inc. (2005). "Takeout Takes Off—Technomic Launches *Tapping the Market Potential for Takeout Dining* Study," Press release, January 27.

Technomic Information Services, Inc. (2006). *2006 Technomic Top 500 Chain Restaurant Report*.

Turner, T. (2005). "Sales Pitch on Aisle Five," Knight Ridder News Service, September 17.

United States Census Bureau. Economic Census. *http://www*.census.gov/econ/census02/.

United States Department of Agriculture. National Agricultural Statistics Service (NASS). http://www.nass.usda.gov/index.asp/

United States Department of Agriculture, Economic Research Service. ERS Briefing Room, "Food CPI, Price and Expenditures: Foodservice as a Share of Food Expenditures"http://www.ers. usda.gov/ Briefing/ CPIFoodAndEx-penditures/Data/table12.htm

United States Department of Agriculture. Grain Inspection, Packers and Stockyards Administration. *Packers and Stockyards Statistical Report*, selected issues.

United States Department of Agriculture. (2004). *Report to Congress on the Economic Effects of U.S. Dairy Policy and Alternative Approaches to Milk Pricing*. July.

Veiders, C. (2005). "Alternative Access: Drug Stores," *Supermarket News*, October 24.

Vocke, G., Allen, E. W. & Ali. M. (2005). *Wheat Backgrounder*, WHS-05k-01. U.S. Department of Agriculture, Economic Research Service, December.

Vranica, S. (2005). "Anywhere, Anytime," *The Wall Street Journal*, November 21.

Zadek, S. (2005). "The Five Steps of Corporate Responsibility," *Harvard Business Week*, January 17.

Zwiebach, E. (2005). "Alternative Access: Supercenters," *Supermarket News*, October 24.

Zwiebach, E. (2004). "Rationalizing the Volume," *Supermarket News*, June 28,.

Zwiebach, E. (2005). "SEC Details Fleming's Accounting 'Illusions'," *Supermarket News*, September 26.

Zwiebach, E. (2004). "Wal-Mart Logistics May Swamp Distribution Chain," *Supermarket News*, April 12.

End Notes

[1] In 2005, Dollar General, the leading dollar store, ranked 18th in U.S. grocery sales ("SN's Top 75," *Supermarket News*, http://www.super-marketnews.com/, accessed October 13, 2005).

[2] A store's private label can be the store's own name or a brand name created exclusively by the retailer for that store.

[3] A study by Global Insight (2005) found evidence that, over the 1985-2004 period, Wal-Mart's investments in innovative distribution and inventory increased the national economy's productivity by helping Wal-Mart suppliers operate more efficiently.

[4] All of Wal-Mart's supercenters and Neighborhood Markets now sell case-ready meats, which Wal-Mart touts as enabling streamlined distribution from manufacturer to customer and providing greater quality control and tracking capabilities. Other factors that have contributed to the growth of case-ready meats in the United States include the move toward branded meat products, retailer preference for more uniform meat cuts, and improved meat-case merchandising without the labor cost of meat cutters.

[5] In addition to its influence over suppliers, other contentious issues raised by Wal-Mart's increasing presence in the economy include low employee wages and shifting of health care benefits onto States as the company pursues a cost-leadership strategy.

[6] The organic market increased by 19 percent in 2004 and is expected to double by 2007 (Tarnowski, 2005; Lempert, 2005).

[7] Some traditional retail outlets, such as Giant Eagle and Raley's, are limiting promotional discounts in favor of everyday low prices, similar to the approach of Wal-Mart and other discount retailers (Adamy, July 2005).

[8] In 2005, mergers between UCCnet and Transora, and between WorldWide Retail Exchange and GNX (referred to as Agentrics LLC), were announced (Garry, June 2005).

[9] One factor driving participation in the GDSN is the emergence of RFID technology and electronic product code, which requires synchronized data (Garry, January 17, 2005).

[10] ERS also calculates grocery concentration ratios on an annual basis, including grocery product sales by Wal-Mart supercenters (Harris et al., 2002). (http://www.ers.usda.gov/Briefing/Food MarketStructures/foodretailing.htm). These ratios are slightly higher than those for grocery stores published by the U.S. Census Bureau, and trends over time are similar.

[11] While national retail concentration ratios provide useful information on the size of the largest food retailers over time, retailers serve consumers in smaller local markets, where concentration can be much higher. ERS analysis of the 100 largest U.S. cities in the 1990s found that increases in local market concentration have been slight (Kaufman, 2000). In the 100 largest U.S. cities, average market share of the four largest food retailers rose from 69 percent in 1992 to 72 percent in 1998.

[12] As with food retailing, the relevant foodservice market is local, and local market concentration can be much higher. Foodservice facilities operate in small-sized markets, such as those in neighborhoods, airports, and malls, and national trends have little to do with how prices are set. While there are over 500,000 foodservice outlets, many operate under a centralized decision process, which could have a role in price-setting behavior.

[13] Ron Marshall, chairman of fifth-ranked food wholesaler Nash Finch, predicted that the number of food wholesalers will likely shrink by more than two-thirds over the next 5 to 7 years and will eventually shrink to a dozen general-line wholesalers serving a national market or multiple regions (American Institute of Food Distribution, Inc., *Food Institute Report,* June 13, 2005).

[14] Mohtadi and Kinsey (2005) note that two types of market structure of the food supply chain likely exist. In one case, the increase in large self-distributing grocery chains, along with other trends, suggest a shift toward retail bargaining power relative to manufacturers. On the other hand, 31 percent of supermarkets are small independent retailers with less than 10 stores, over half of which own only 1 store. In the latter case, market power likely remains with the manufacturer.

[15] Ollinger et al. (2006) analyzed mergers and acquisitions over the 1977-82 and 1982-87 periods in specific manufacturing industries, including the meat and poultry slaughter and processing, dairy, flour, feed, and oilseeds industries. They found that acquired plants were highly productive and became even more productive after the merger. The 1997-2001 period was an especially active time for mergers and acquisitions by food manufacturers

(see the section "Recent Mergers, Acquisitions, and Divestitures in the Food Industry," p. 27).

[16] Private market research companies, such as A.C. Nielsen and Information Resources, Inc., provide more narrowly defined product market definitions (Conner and Schiek, 1997). For example, while NAICS has one classification for the breakfast cereal industry, private companies distinguish between ready-to-eat and cooked breakfast cereal.

[17] The average CR4 excludes eight local markets and "all other miscellaneous food." For industries serving localized markets, such as fluid milk, ice cream, ice, bakeries, animal feed, soft drinks, and snack foods, local concentration is likely much higher. For the category "all other miscellaneous food," the CR4 is meaningless.

[18] ERS research found size economies to exist in hog production, even for the industrial-sized hog farms (5,000 or more hogs) (McBride and Key, 2003).

[19] Several ERS studies have analyzed the growth in hog contracting. These studies have found that through contracts, packers can gain greater control over hog quality attributes and ensure a steady supply of uniform animals, while producers are assured of a market outlet. Policy concerns include volatile and thinly traded spot markets, proprietary nature of contract information, and the ability of smaller, independent producers to compete with contract producers (Martinez, 1999; Martinez and Zering, 2004).

[20] Using data from 1963 to 1992, ERS research found that economies of scale were one factor that led to consolidation in hog and cattle slaughter (MacDonald et al., 1999).

[21] According to analysis by MacDonald and Ollinger (2005), the rate of cost reductions related to consolidation in the beef industry slowed after 1992 because of reduced consolidation and smaller cost advantages of the largest sized plants.

[22] Local independent and regional retail grocers may join together to own and control wholesale establishments, providing scale economies to compete with the large retail chains. In addition to buying goods at lower costs, these grocery wholesale cooperatives provide other services, such as advertising, technology, and risk management. They rank among the largest cooperatives, food and nonfood, and grocery wholesalers in the Nation. Those included among the top 10 grocery wholesalers are Wakefern, Associated Wholesale Grocers, and Unified Western Grocers (Appendix A, Appendix table 1).

[23] Marketing Intelligence Service's Productscan Online is a worldwide new product database that includes content from two Marketing Intelligence Service, Ltd., publications: *Product Alert* and *International Product Alert.*

[24] In 2005, Marketing Intelligence Service's Productscan Online data base listed a total of 100 U.S. food and beverage product claims or tags.

[25] Diets, such as the Atkins, usually exhibit sensational ups and downs, but come and go afterwards, rather than totally disappearing, as a true fad does (Knowledge@Wharton, 2005).

[26] Detailed Census information is available at 5-year intervals

CHAPTER SOURCES

The following chapters have been previously published:

Chapter 1 – This is an edited, reformatted United States Department of Agriculture Agricultural Marketing Services Program Handbook 728-3, dated July 2008.

Chapter 2 – This is an edited, reformatted California Institute for Rural Studies, dated November 2009.

Chapter 3 – This is an edited, reformatted United States Department of Agriculture Economic Research Report Number 42, dated May 2007.

INDEX

A

accessibility, 11
accounting, 18, 27, 81, 100, 118, 120, 143
acquisitions, 114, 121, 128, 134, 135, 144, 146, 147, 148, 173
advocacy, 90
aesthetics, 85
affluence, 19
agencies, 81, 99, 100, 103
aggregation, 89, 104
agility, 33
agricultural market, 88
agriculture, 11, 25, 43, 51, 57, 61, 64, 74, 76, 88, 103, 105, 108, 139, 144
airports, 173
animal husbandry, 30
animal welfare, 152
annual rate, 131
antitrust, 131
anxiety, 44
apples, 123
assessment, 138
assets, 30, 130, 146, 147
audits, 86
authenticity, 49
average costs, 133

B

bankruptcy, 120, 130, 148, 157
bargaining, 136, 173
barriers, 44, 57, 60, 61, 62, 63, 74, 75, 82, 83, 84, 91, 93, 98
beef, 29, 43, 126, 144, 147, 150, 174
beer, 131, 165
Belgium, 131
beverages, 18, 19, 118, 153, 156
birds, 29
brand loyalty, 28, 154
breeding, 145
bun, 123
business management, 25
business model, 33
Butcher, 126
buyer, 26, 27, 34, 44, 57, 58, 62, 71, 72, 84, 85, 86, 87, 88, 91
by-products, 29

C

cacao, 163
calorie, 123
campaigns, 59, 81, 100, 111
cane sugar, 138, 140
capital intensive, 96
carbohydrate, 123, 141, 157
case study, 35
catalyst, 41
cattle, 138, 145, 146, 174
Census, 11, 13, 113, 114, 115, 125, 134, 136, 137, 138, 139, 140, 141, 143, 144, 146, 163, 164, 165, 171, 173, 174

certification, 30, 44, 47, 58, 59, 66, 69, 72, 77, 78, 79, 81, 86, 96, 98, 99, 100, 103, 108
challenges, 56, 57, 58, 59, 60, 61, 62, 63, 64, 66, 70, 71, 72, 73, 74, 77, 81, 82, 83, 85, 86, 88, 90, 91, 92, 99, 104, 107
character, 160
cheese, 27
chicken, 123, 147
China, 75
City, 22, 160, 164
clarity, 78
classification, 165, 166, 173
cleaning, 5, 27
clients, 48
climate, 2
close relationships, 150
coatings, 123
coding, 165
coffee, 124, 127, 131, 159
color, iv
commercials, 157
commodity, 61
communication, 27, 33, 34, 35, 58, 80, 82, 87, 93, 99, 102, 130
community, 11, 36, 70, 80, 93, 120
competition, viii, 3, 4, 56, 57, 58, 67, 71, 72, 75, 79, 86, 95, 111, 112, 113, 116, 123, 125, 126, 128, 131, 135, 148
competitive advantage, 23, 25, 116, 117, 125, 130, 151
competitors, 14, 23, 28, 30, 48, 118, 130, 135, 153
complement, 59
compliance, 76, 151
computer skills, 82
computer systems, 128
computerization, 35
concentration ratios, 146, 173
conference, 169
conservation, 103
consolidation, 3, 46, 58, 59, 74, 88, 102, 104, 112, 114, 128, 133, 134, 135, 137, 143, 144, 145, 146, 174
consulting, 52
consumer demand, 7, 12, 17, 18, 20, 32, 33, 42, 48, 56, 60, 61, 75, 89, 102, 104, 123, 149, 157
consumer education, 81, 99, 100, 105
consumer expenditure, 10, 112
consumer loyalty, 4
consumer markets, 51
consumption, 5, 9, 13, 108, 114, 131, 140, 144, 149, 154
consumption patterns, 108, 149
contamination, 47
convention, 50
cooking, 127, 139, 158
cooling, 85
coordination, 33, 34, 114, 137, 146
cost, 2, 32, 33, 34, 35, 36, 41, 42, 44, 48, 50, 75, 78, 82, 84, 89, 90, 100, 103, 112, 113, 117, 125, 133, 135, 144, 151, 172, 174
cost of living, 75
cost saving, 32, 36, 133, 135
costs of production, 75
criticism, 150
critics, 151
crops, 50, 57, 61, 62, 68, 70, 72, 73, 86, 93, 95
culture, 159
customer preferences, 7, 33, 120

D

dairies, 43
dairy industry, 116, 143
damages, iv
data collection, 49, 103, 153
database, 59, 90, 103, 109, 121, 129, 174
decentralization, 50
defects, 85
demonstrations, 44
Department of Agriculture, v, 1, 108, 109, 123, 144, 168, 169, 170, 171, 172
Department of Commerce, 132, 133, 134, 136, 139, 140
destination, 46, 47
dichotomy, 78

diet, 123, 155
digital communication, 82
direct investment, 131
directors, 58, 63
disappointment, 42
disclosure, 150
District of Columbia, 22
divergence, 15
diversity, 59, 95, 114, 130
dominance, 4
dough, 162
dressings, 22
drugs, 5
dynamics, 108

E

earnings, 41, 96
e-commerce, 128
economic activity, 114, 133
economies of scale, 90, 133, 135, 174
economy, 12, 31, 116, 172
Education, 105, 106, 107
employees, 40, 122, 139, 149
employment, 94, 114, 122, 149
enforcement, 60, 96, 100
entrepreneurs, 43
environmental issues, 49, 150
environmental movement, 64
environmental standards, 75
environmental sustainability, 29
EPC, 36, 40, 121
equipment, 33, 35, 86
equity, 16
erosion, 4
ethnic groups, 15, 18
ethnicity, 65, 73
expenditures, 5, 114, 118, 119, 123, 149, 154, 156
expertise, 62, 126, 130
experts, 57
exports, 116, 130, 131, 132, 139
exposure, 23, 148

F

family farms, 43, 78
farm size, 57, 59, 62, 72, 81, 100
farmland, 56, 60, 66
farms, 12, 23, 43, 57, 58, 59, 60, 61, 62, 63, 65, 66, 67, 71, 72, 74, 75, 76, 77, 78, 79, 80, 81, 82, 83, 84, 85, 86, 87, 88, 89, 90, 91, 92, 95, 98, 99, 100, 101, 102, 103, 104, 108, 109, 115, 143, 145, 173
FAS, 131
fast food, 123
fat, 123, 157, 158
fatty acids, 158
fiber, 114
films, 157
financial performance, 135
firm size, 34
fishing, 51
flavor, 12, 42, 153
flour, 131, 140, 162, 173
fluid, 140, 143, 144, 173
focus groups, 63, 64
food industry, 32, 33, 36, 41, 105, 112, 114, 116, 118, 120, 157, 165
food prices, 14, 112, 135
food production, 28
food products, vii, 1, 3, 5, 19, 22, 27, 29, 31, 32, 47, 49, 50, 65, 112, 116, 126, 131, 149, 159, 160
food safety, 47, 49, 58, 59, 76, 81, 83, 86, 98, 99, 100, 103, 105
forecasting, 36, 40, 41
foreign direct investment, 116, 131
formula, 4
fraud, 96
frost, 86
fruits, 26, 57, 61, 68, 121, 131
funding, vii, 55, 81, 103

G

Galileo, 171
Georgia, 51
Germany, 171

global economy, 33
global trade, 129
globalization, 114, 130
goat milk, 43
goods and services, 20
governance, 106
grading, 27
grant programs, 96
grass, 43
gross domestic product, 114
growth hormone, 18
growth rate, 14, 15, 17, 49
guidelines, 151, 152

H

harvesting, 93
health insurance, 94
host, 123
hub, 88
human rights, 153

I

ideal, 138
image, 7, 113, 126, 153, 160
immigrants, 25
Impact Assessment, 105
impacts, 67
imports, 58, 75, 79, 116, 131, 132
incidence, 35
increased access, 130, 134
increased competition, 77, 135
independence, 159
individuality, 159
industrialization, 144
inefficiency, 95
inferiority, 27
inflation, 14
information exchange, 34
information technology, 32, 34, 128, 129
insight, 131
inspectors, 47
integration, 2, 33, 34
intermediaries, 58, 65, 76, 87, 89, 92, 109
Internal Revenue Service, 154, 168
international trade, 131
investors, 146, 149
isolation, 33
issues, 36, 37, 57, 58, 62, 63, 71, 72, 73, 74, 77, 78, 85, 99, 103, 108, 150, 152, 166, 172
Italy, 106

L

labeling, 30, 60, 79, 81, 96, 101, 107, 123
landscape, vii, 1, 2, 66, 77
language barrier, 58, 59, 92, 93
Latinos, 25
leadership, 117, 172
learning, 18, 33, 61, 86, 97
legislation, 96
lending, 88
liability insurance, 26, 86
linen, 127
Lion, 126, 161
litigation, 151
livestock, 68, 97
local government, 96
logistics, 25, 32, 34, 36, 40, 41, 83, 84, 120, 130
lower prices, 28, 84, 117, 120
loyalty, 128

M

magazines, 80, 101, 113, 119, 156
majority, 5, 7, 61, 68, 69
management, 32, 35, 91, 108, 128, 129
mandates, 121, 150
manufacturing, 36, 111, 114, 121, 130, 131, 132, 137, 138, 139, 140, 152, 164, 165, 173
manufacturing companies, 131, 152
market access, 42, 44, 104
market concentration, 133, 173
market segment, 7, 125
market share, 17, 30, 116, 123, 125, 136, 144, 153, 173

market structure, 114, 133, 144, 173
marketing strategy, 3, 79
marketplace, vii, 1, 13, 21, 31, 32, 44, 49, 77, 92
Mars, 154
meat, 5, 17, 29, 76, 97, 131, 144, 145, 146, 155, 163, 172, 173
media, 13, 150, 151, 156, 160
median, 68
merchandise, vii, 1, 3, 4, 6, 8, 9, 23, 27, 28, 31, 43, 48, 50, 114, 117, 120, 121, 122, 127
mergers, 113, 114, 128, 134, 135, 137, 144, 146, 147, 148, 173
messages, 87, 157
Mexico, 75, 89
mixing, 163
Moldova, 76
momentum, 64
motivation, 18
multiples, 6

N

natural food, 4, 6, 19, 26, 30, 48, 60, 65, 77, 125, 126
negative outcomes, 150
Netherlands, 131
networking, 80, 100, 102, 103, 105
NGOs, 150
niche market, 74, 77, 137
nickel, 44
North America, 106, 107, 124, 129, 134, 144, 146, 161, 165
nutrition, 7, 49, 114, 123, 152

O

obesity, 151
obstacles, 87
oil, 131, 139, 158, 159
oilseed, 139, 163
opportunities, vii, 1, 18, 21, 32, 49, 60, 64, 72, 83, 90, 92, 96, 98, 100, 102, 103, 104, 105, 108, 117, 131, 159
organ, 38
organic food, 6, 12, 16, 17, 18, 48, 56, 60, 64, 65, 66, 67, 82, 84, 87, 103, 104, 105, 107, 109, 126, 127
output per worker, 135, 137
outreach, 77, 82, 98, 99, 103, 104
oversight, 36
ownership, 137

P

Pacific, 43
pairing, 83
paradigm, 2
pasta, 142, 162
pasture, 43
performance, 47, 52, 112, 117, 146, 152
performers, 14
permission, iv
permit, 28
personal relations, 12, 58, 74, 80, 104
personal relationship, 12, 58, 74, 80, 104
physiology, 26
plants, 29, 130, 131, 137, 138, 139, 140, 141, 142, 143, 144, 145, 146, 162, 163, 173, 174
platform, 113, 129
polarization, 15
pools, 34, 36, 129
population growth, 14, 19, 51
portfolio, 39, 40
positive feedback, 95
potato, 24
poultry, 5, 17, 29, 131, 162, 173
price competition, 135
price deflator, 138
price index, 14, 143
procurement, 2, 32, 36, 40, 41, 42, 47
producers, vii, 1, 3, 18, 23, 25, 27, 28, 32, 34, 42, 43, 44, 48, 49, 50, 63, 66, 74, 75, 76, 88, 89, 97, 106, 107, 126, 145, 146, 147, 174
product attributes, 49
product market, 18, 144, 173
product performance, 42

production capacity, 26
production costs, 75, 84, 98, 133
profit, 16, 32, 44, 120
profit margin, 32, 44
profitability, 11, 30
project, 97, 113, 126
proliferation, 92
proposition, 79
public policy, 151
pumps, 127
purchasing power, 19, 49

Q

quality assurance, 138
quality control, 2, 3, 28, 32, 47, 138, 172
quality improvement, 138
quality standards, 19, 27, 44, 85

R

radio, 3, 121, 156
real estate, 154
real-time basis, 36
recall, 28, 29
recommendations, iv, 57, 59, 61, 63, 64, 81, 90, 96, 98, 99, 102, 105, 123
Registry, 129
relaxation, 127
relevance, 98, 102
reliability, 44
repackaging, 46
reputation, 2, 27, 28, 151
requirements, vii, 1, 2, 3, 40, 46, 47, 48, 57, 59, 71, 72, 86, 91, 100, 137
resources, 50, 77, 101, 122
respect, 56, 58, 93
response time, 33, 37
responsiveness, 149, 157
restaurants, 65, 69, 73, 86, 91, 108, 109, 112, 123, 125
restructuring, 128, 130, 148
revenue, 30, 44, 121, 124, 148
risk management, 95, 174
royalty, 160
rural areas, 92

S

saturation, 94
savings, 28, 32, 39, 40, 41, 135
scale economies, 144, 174
scarcity, 67
seafood, 29, 140, 143, 162
selectivity, 107
self-image, 113
semi-structured interviews, 57
service provider, 46, 86, 88
shareholders, 94, 149
SIC, 134, 136, 138, 165
signs, 135
small businesses, 18, 118
social interests, 2
social justice, 79
social network, 80, 82
social responsibility, 111, 114, 149, 150, 153
social welfare, 30, 151
sodium, 123
software, 3, 33, 35
Southeast Asia, 25, 78, 109
soybeans, 159
space, 6, 46, 88, 94, 127, 153
specifications, 7, 29
speech, 51
spot market, 114, 146, 174
stakeholders, 150
statistics, 13
storage, 26, 27, 85
storytelling, 82
strategy, 28, 58, 79, 95, 97, 113, 117, 123, 124, 125, 130, 131, 133, 144, 149, 153, 160, 172
structural changes, 112, 143, 144, 145
substitutes, 138
supply chain, 2, 3, 32, 33, 34, 35, 36, 38, 41, 49, 62, 63, 65, 87, 88, 104, 113, 120, 128, 150, 154, 159, 173
survey, viii, 7, 8, 13, 32, 51, 55, 57, 59, 62, 64, 68, 69, 70, 71, 72, 73, 75, 81, 82, 90,

91, 96, 97, 98, 99, 101, 127, 128, 129, 130
survey design, viii, 55
sustainability, 106, 150, 151
sustainable growth, 149
Sweden, 169
synchronization, 32, 35, 36, 38, 40, 41, 49, 129

T

tactics, 113
tags, 34, 38, 121, 130, 157, 158, 174
technical assistance, 91, 101, 103, 104
technological progress, 153
temperature, 27
tension, 2
territory, 42
testing, 120
time commitment, 94
tobacco, 22, 48, 119, 121, 122, 131, 154, 165
tourism, 69, 80, 97, 101
toys, 127
tracks, 47
trade agreement, 130
trade deficit, 131
trading partner, 36, 40, 129
trading partners, 36, 40, 129
traditions, 12
training, 25, 150
transaction costs, 58, 83, 84, 102, 104
transition period, 99
transmission, 49, 50
transportation, 25, 37, 41, 44, 46, 93, 130, 131
trends, 3, 5, 14, 18, 30, 139, 140, 145, 173

U

U.S. history, 9
uniform, 38, 86, 172, 174
universities, 92, 103
updating, 27
urban areas, 92
USDA, vii, 4, 5, 11, 23, 24, 31, 50, 55, 56, 66, 75, 76, 78, 92, 97, 105, 106, 107, 108, 109, 115, 131, 132, 143, 144, 145, 146

V

variations, 117, 153, 158
vegetables, 26, 57, 61, 68, 73, 121, 131, 155
venue, 13
vertical integration, 3, 114, 149
video, 113, 127, 154, 156, 157
video games, 113, 156, 157
videos, 127
voting, 15

W

wages, 172
waste, 90, 94, 95
wear, 76
wellness, 16
wholesale, 2, 4, 5, 6, 57, 59, 62, 65, 69, 71, 72, 73, 77, 82, 83, 84, 85, 86, 87, 90, 91, 94, 95, 98, 99, 101, 102, 104, 108, 109, 121, 125, 127, 128, 129, 130, 134, 164, 165, 174
wireless technology, 3
workers, 19, 94